TOWARD A MATHEMATICS OF POLITICS

Toward a Mathematics of Politics

by Gordon Tullock

Ann Arbor Paperbacks
The University of Michigan Press

First edition as an Ann Arbor Paperback 1972

ISBN 0-472-06187-9 (paperback)
ISBN 0-472-92600-4 (clothbound)
Published in the United States of America by
The University of Michigan Press and simultaneously
in Don Mills, Canada, by Longman Canada Limited
Manufactured in the United States of America

Preface

In the last fifteen years or so a body of literature applying essentially economic tools to the analysis of politics has developed. In addition to a considerable number of articles and monographs, there are seven or eight books in the field. But though all of this work is recognizably "economic" in its methods, the models used are in many ways quite different from those in economics. Further, as would be anticipated in a new field of investigation, there are subtle inconsistencies among the models. A unified model which eliminated these minor, but irritating, inconsistencies and dealt with both economic and political problems would be highly desirable. That it is desirable does not, of course, prove that it can be produced immediately. Physics has not achieved its general field theory and it is much older. Nevertheless, the National Science Foundation thought enough of the prospects to give me a research grant to look for such a unified theory. In my application for the grant I proposed to search for the theory, I did not promise to find it. This is fortunate, because I have not found the grail, although I can truthfully say that I have devoted much time to the quest and had many interesting intellectual adventures on the way.

This book, in a sense, is the account of my search. It reports various things that I have discovered while trying to reach my elusive goal. To quote Mao Tse-tung, it is "the first step on the 10,000 Li march." I tried a number of approaches, some of which led to interesting results even though they did not lead to a general theory. The chapters of this book recount these intermediate results. I intend to continue the search, and I hope that other people will join me. Meanwhile, some of my results should be of use to scholars in both economics and political science whose main objectives are different from my own. Science is always a march toward an objective which we never reach because we are approaching it

asymptotically. Still, the reader should not expect to find a single theme carefully worked out in various chapters.

A brief breakdown of the chapters may be of assistance. Chapters II, III, and IV constitute an application of a geometrical tool invented by Duncan Black to various political problems. Chapters V and VI use another of Black's inventions to deal, first, with the economic problem of monopolistic competition and, then, with an investigation of the pattern of information flows to be expected when the provision of information is organized on market lines. Chapters VII, VIII, and IX deal with political information, elaborating on a theme by Anthony Downs. The first and last chapters each deal with a separate theme. Although there is a good deal of difference in the subject matter and approach of the chapters, they are not unrelated nor are they arranged in a random manner. The reader will find the reasoning easiest to follow if he reads them in the order in which they occur. The book is primarily aimed at students of the new political theory, but it should also interest specialists in persuasion and propaganda. Such people will find most of the matter of direct interest to them in Chapters I, VI, VII, VIII, and IX.

An article based on Chapter III was published in the *Quarterly Journal of Economics,*[1] and one based on the first portion of Chapter V appeared in *The Western Economic Journal.*[2] I am grateful for permission to republish this material. I must also express my gratitude to Kenneth Arrow, James Buchanan, David Chapman, John Moore, William Riker, and Oliver Williamson, all of whom made useful suggestions for improvements. It is customary, at this point, to say that any errors remaining in the manuscript are to be blamed solely on the author. I have always wondered (even when conforming) at this custom. Certainly, it is conceivable that someone gave me some bad advice. Also, with six such distinguished experts reading the manuscript in whole or in part, it would seem that any defects would have been noticed. Thus it seems to me at least possible that any errors remaining in the book might be their fault instead of mine. Mrs. Betty Tillman gave her usual competent secretarial assistance.

1. "The General Irrelevance of the General Impossibility Theorem," (May 1967), p. 256.
2. "Optimality with Monopolistic Competition" (Fall 1965), p. 41.

Contents

CHAPTER I

Models of Man

In modern economics and in the political theory which is now developing out of economics, the preference schedule has substituted for the man. Researchers have not engaged in elaborate speculation about the nature of man or the reasons for an individual's desire of some certain thing. We observe that different people want different things, and that the same person will want different things at different times. Further, it is a matter of common knowledge that people frequently want things which, in the view of other persons, they should avoid. In other words, people have preferences with respect to the behavior of others. We may be made unhappy by the choices of other people. My neighbor may choose to paint his house bright purple, lead a sex life which I find offensive, treat his children with conspicuous cruelty, or take up the manufacture of nitro-glycerin in his basement. All of these things would be expressions of his preferences, and my distaste for them would be an expression of mine.

My neighbor might, on the other hand, do things of which I approve. He might, for example, plant his garden so that the view from my house is improved. He might do this from vanity, a genuine desire to give his neighbors pleasure, or simply as an accidental by-product of following his own tastes. St. Peter, presumably, will have to consider the question of motives, but the economist traditionally merely notes what my neighbor's preferences are and does not inquire as to the reasons for them. Similarly, my neighbor may reduce his personal consumption to make gifts to the poor, rigidly follow a moral code which I believe good, or lose his life while saving a child in a burning building. Commendable as all of these actions may be, to the economist they have been simply expressions of the man's preferences. For the psychologists these preferences are problems for investigation, but the economist has normally accepted them as givens upon which his analysis may be based.

It is not my purpose to criticize the economist for leaving the factors which form preferences out of account in his analysis. A great deal of useful research has been done using this model of man, and a great deal more has yet to be done. The ancient economic principle of the division of labor, however, indicates that we should not all do the same thing. It is sensible to have some economists investigating the foundations of individual preferences. Advertising is an economic activity, and certainly at least some of it is directed toward altering people's preferences.[1] A system which takes preferences as given can hardly analyze efforts to change them. Thus, even in the strictly economic field a system of analysis which is not based on "revealed preference" would be of considerable use.

In politics the case for considering factors effecting preferences is even stronger since much political activity is directed toward that end. The educational machine, so important in almost all modern governments, has as one of its most important objectives the instilling of a set of "desirable" preferences in the population. We hope that it will turn out citizens who like good literature, are patriotic, and recognize that their neighbor's property is, indeed, their neighbor's. The frequency with which these hopes are disappointed simply leads to greater educational efforts. But, even if we leave the educational system aside, efforts to change preferences are a much more conspicuous feature of our political system than of our economic organization.[2]

Needless to say, politics does not entirely turn on attempts to alter preferences. As in economics, models in which preference orderings are taken as given will explain much political behavior. Such models, in fact, will be used a good deal in this book. The need, however, to supplement these models by others in which preferences are shifted is greater in politics than in economics. If I find that my neighbor's cruelty to his children annoys me, I may try by persuasion to change his preference schedule so that he no longer wishes to beat the child. If this cause has enough political support, the state may use its propaganda facilities to reach the same end. Any discussion of this situation will require a model in which preferences are consequences, rather than givens.

More commonly, however, the state would deal with this problem by making arrangements which change the alternatives among which my neighbor must choose. By making excessive child beating a crime, it makes my neighbor choose between two package alternatives: he can beat his child and go to prison or refrain and stay free.[3] If his dislike of prison exceeds his desire to beat his child, the child will remain unbeaten. This requires no change in his preferences, and a model in which preferences are assumed as basic data is ideal for discussing such problems. A majority, probably a large majority, of political problems are of this sort, and hence may be dealt with by models in which the preference ordering substitutes for the man.

Genuine changes in preferences are, however, made or attempted in both the political and economic fields, and to deal with them we need a little more elaborate view of the choosing process than is utilized in conventional economics. The bulk of this chapter will be devoted to developing a somewhat more complex preference mechanism. This "theory" will be basically rather similar to the existing economic idea of preference,[4] and will, like the conventional theory, rest upon relatively obvious assumptions. Like the presently used assumptions on preference, which, indeed, will be simply a special case of our more general system, our model of preference will be more or less empty of psychological content. Neither system will be of much use to the psychologists. We will continue to ignore the basic reasons why people prefer things, and simply note that they do have a preference structure. The system will be useful for the study of social interaction, but not for the study of personality.

First, however, a brief diversion into the problem of transitivity. Two separate assumptions about preferences have been used by modern economists. One of these is simply that the individual orders all alternatives, and that the schedule produced is his total preference schedule. The second is that he will be able to make choices among pairs of alternatives, unless he is indifferent between them. The second assumption, of course, is deducible from the first. Instead of making this deduction, however, the trend of modern writing in the field has been to assume it independently, probably because it is an obvious and readily testable statement about the real world. From this assumption and a further assumption, that such choices are transitive, it is possible to deduce the preference schedule, and most modern economists have taken this route.

The assumption that preferences are transitive, that if A is preferred to B and B to C then A will be preferred to C, has been questioned. Logically it would be possible for an individual to choose A over B, B over C, and C over A. If he did it very often we would have him put under psychiatric supervision, but this has to do with our views of human nature, not with a logical inconsistency between his behavior and the basic axiom of choice. It is rather unfortunate that the assumption that an individual is able to choose was put in the form of a pairwise comparison rather than as the ability to choose the "best" alternative from among several. The latter assumption, which is just as obvious and easily testable as is the choice from two, rules out simple intransitivity without the need of a special assumption on the point.[5]

There is, however, a real problem which is not entirely one of pure logic. Experimental work does seem to show occasional intransitivity among paired choices. Until very recently it was my opinion that these apparent intransitivities were the result of the fact that the experiments necessarily took time, and that the subject changed his mind during the course of the experiment.[6] I still think that most of them arise from this cause, but I am now willing to admit that genuine intransitivities may exist. They are, essentially, intellectual errors. One of the features of the decision-making model to be presented is that it is possible for an individual to make an error about his own preferences.

Note, however, that this will make no particular difference to traditional economics. It has always been accepted that choices made at different times may be somewhat different. People like variety, and they do change their minds. These "inconsistencies," however, are normally fairly small. The model individual in classical economics when engaging in action where he must take into account the probable decisions of others, does not assume that these decisions are unchanging, but only that they change slowly. He need not concern himself with the question of the origin of the changes. Whether the individual's preference schedule contains intransitivities or is simply changing slowly is an unreal question for the practical politician or the merchant. Economists have sometimes assumed that these preferences do not change at all, but this is a simplifying assumption which makes the reasoning easier, not a statement about the real world. As a simplification, it is most useful and we shall make considerable use of it, but neither conventional economics nor our analysis really depends upon it.

It is sometimes said that we should consider only "revealed preferences," choices that people actually make between real alternatives. It is frequently wise to confine our investigation to such choices, but we should avoid equating these rules for our investigation with statements about preferences themselves. It is true that we usually know no more than that some individual has made certain choices, and it is sensible to consider situations where such choices determine a set of ordinal preferences. One of the great achievements of modern economics is to demonstrate that cardinality of preferences is not necessary for most purposes. In general, all that is needed is a monotonic series. It is not true, however, that the individual's preference pattern is as unstructured as would appear from the ordinal schedule which we use to represent it. The individual may not be able to accurately describe his preferences in terms which permit us to make cardinal computations, but we can be fairly sure that this is a problem of limits on information. He himself does make cardinal computations and decides whether he wants various things at various prices in a much more sophisticated manner than is indicated by his "revealed preferences."

We do know more about human preferences than that they are an ordinal list, and this fact is implicitly used in certain parts of economics. All economists will agree that, with certain specified exceptions, individuals will not sacrifice as much to get a second item of a certain sort as to get the first. They will normally be quite happy to make predictions of this type with respect to a completely unknown man choosing among alternatives unknown to them. As another example, it will be generally admitted that the composition of the choices made, changes as the budget constraint is raised. A man who must choose five items from some given collection will not, if given an opportunity to choose ten, simply choose twice as much of each item as he took the first time. Neither of these predictions about human behavior can be derived from a simple ordinal list of preferences. Since these predictions would be accepted by every economist, it follows that the economist actually has a view of human nature which is more complex than the simple ordinal preference schedule.

There is another normal, everyday phenomenon which cannot be explained in terms of schedule of revealed preferences. It is common for completely new products, services, or political platforms to be created. The people who think up these innovations clearly try to produce something which will be preferred to the presently existing alternatives. They must have some idea of how people will act when confronted with a choice they have never previously encountered. If we look around us, we see that this particular form of activity, the creation of new alternatives, is prosperous and expanding. This can only be explained on the assumption that people have some mechanism for choosing which will provide choices in new situations, and that it is possible to make fairly accurate guesses as to what such choices would be. In order to explain this very common economic and political activity, we need a more complex model of the human choice mechanism.

There is, unfortunately, a widely held myth about how new products and services in the economic field are introduced. It is sometimes alleged that these things are not really needed, but that the desire for them is created by advertising. For some reason this allegation is seldom made about new political ideas. The rise of democracy is not explained as the result of shrewd men taking people who were basically satisfied with despotisms, and hypnotizing them into demanding a new form of government. Nevertheless, if new desires can be created in the economic field, clearly the same would be true of the political. Fortunately the myth is a myth. It is obviously true that advertising does affect consumption, and, as we shall see, it is probably true that it does change desires, but this is a secondary and minor effect. If new wants could be created at will by advertising, then companies would waste no money on laboratories and research. Any product, chosen at random, could be sold by advertising. A cure for cancer and a small culture of gangrene bacteria would have about the same sales potential, and the only significant variable would be the respective advertising campaigns.

This is, of course, absurd. Even if we grant that advertising can sell anything, which is clearly not true,[7] the advertising costs of selling some products would be higher than the costs of selling others. The businessman would want to economize and, consequently, would choose products for which the likely advertising cost was small. The cure for cancer, for example, would need almost no advertising and hence would be a highly profitable item compared to the gangrene bacteria. In the real world advertising campaigns frequently fail entirely. The people are not as stupid as the proponents of the "created wants" myth believe. Penicillin and the Volkswagen were both tremendously successful although they were initially marketed without advertising. The complete failure of the Edsel provides the contrary case. The businessman or politician, if he is wise, will take the preferences of the people into account. If he assumes they are infinitely plastic, that advertising can do anything, he will find that his more perceptive competitors always win.

If we examine the skeletons of the men of the late Stone Age they do not appear much different from those of modern man.

Most anthropologists do not think that much change has occurred in the basic physical and mental equipment of human beings since that date. Men have, however, created a vast number of things which were not available in the Stone Age, and now consume them in great quantity. If we accept the anthropologist's view that we are much the same as the men of the old Stone Age, we must explain how these Stone Age men somehow had potential demands for Rembrandt's, Cadillac's, and cancer cures. They can hardly have been equipped with a simple preference system which only distinguished among things in their own environment, since they proceeded to invent and use things which were not in that environment.

What we need, then, is a systematic model of preferences which will explain the existence of potential preferences for things that do not now exist, and which can accommodate itself to the existence in the real world of efforts to make at least some changes in preferences. We would also like the model to pass what can be called the "evolutionary" test. As far as we can tell, no basic changes in men's mental equipment have been made since the end of the Stone Age. If we believe in evolution, we must believe that the human mind evolved under Stone Age and pre-Stone Age conditions. We should not, therefore, put anything in our model which would be desirable equipment for modern man unless it would also be of use to a cave man. Clearly, a simple ordinal list will not do. There is no evolutionary reason why a cave man should have had a preference for El Greco's view of Toledo over a color photograph of the British royal family.

Instead of assuming that we have built-in schedules upon which all existing or potential objects are listed in order of preference, let us think of the individual having "wants." These "wants" are not specific, but specific objects and services fulfill them to varying degrees. Exactly what these wants are, we do not know, but the psychologists may eventually find the answer. Fortunately, a very general model of preference can be developed without such specific information. For this purpose we can consider preferences with respect to the physical world. As an example, we can consider some person's preferences with respect to coffee. He likes it a certain strength, with two lumps of sugar and cream. He also prefers some particular brand. We do not need to inquire what wants these various aspects fill, we can simply discuss his relative preferences for physical variations in the "recipe."

This distinction between "wants" which are part of the internal mental apparatus of the man and his preferences with respect to the real world is fairly readily explained by use of an example. Assume that some individual has as his internal "want" system a drive to maximize an expression like f (a,b,c,d,e,f,g). We do not know what a, b, etc., are, nor do we know the shape of the function relating them. We can, however, determine their effect on real world choices. Suppose we hold everything about the coffee constant but vary the amount of sugar in it. If we plot the amount of sugar on the horizontal axis and the desirability of the resulting product to some individual on the vertical axis, we would expect to get something rather like the line in Figure I.[8] This line does not necessarily represent any one of the arguments in the want function. It is simply the result of that function plotted along a variable which may be one of the arguments of the function or effect several of them.

If the individual is given a choice, he would then adjust the amount of sugar he puts in his coffee to reach the maximum. In practice people seem to think that the cost of exact measurement is greater than the desirability of hitting their exact maximum, and take their sugar in lumps or in spoonfuls, without attempting to make finer discriminations. Coffee, of course, varies in many ways other than in the amount of sugar it contains. Suppose that we substitute some other variable for the sugar on the horizontal axis of Figure I, a variable which is controlled by the manufacturer rather than the customer. The manufacturer normally can make infinitely fine adjustments if he wishes, without any difficulty. On the other hand, he can hardly put out a separate brand of coffee for each customer. Here again the costs would be excessive. As a result, the customer will find himself choosing among a finite number of brands of coffee on the market. Let us assume that the brands on the market are W, X, Y, and Z. If we assume that these brands are identical in everything except the variable shown on Figure I, our individual customer would be consuming brand W. Potentially superior alternatives do exist, however, and the manufacturer who offers one will obtain this man's custom.

Consider a manufacturer thinking of entering this one-man marketplace. He is contemplating a product at either A or B. Let us again grant that advertising can do anything, and that it would get our hypothetical individual to choose A instead of the W which he is now consuming. It should, however, be obvious that he could be switched to B with considerably fewer resources in advertising

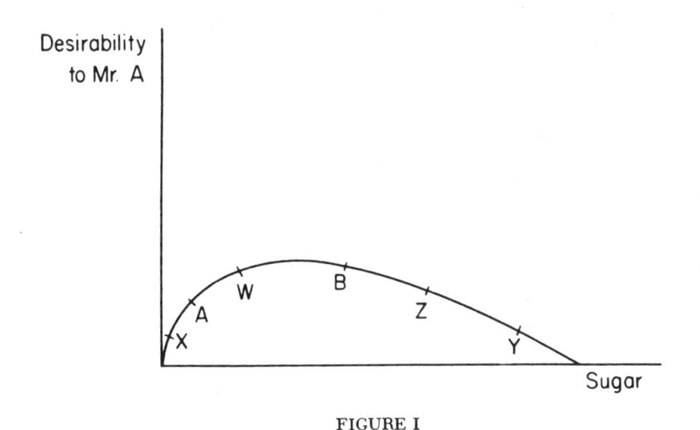

FIGURE I

thrown at him. Assuming A and B can be produced for the same cost,[9] clearly it would be more profitable to introduce B rather than A. The manufacturer who wishes to make as much money as possible must take the preferences of the customer into account even if he thinks that advertising can sell anything. Thus, even granting the "hidden persuaders" myth full credence, it remains true that the manufacturer must serve the customer's convenience.

If we consider a number of different individuals, clearly we could similarly plot their preferences on Figure 1 as a series of curves. If the collection were single peaked, as Black has demonstrated,[10] this would be a suitable situation for majority voting. Economically, however, it is also most convenient. A manufacturer thinking of introducing a new product would, if he knew the shapes of all of these curves, be able to choose the new product so as to maximize his returns. Such perfect knowledge is, of course, not likely, but fairly good knowledge is. If the potential producer simply knows what products are now being produced, and their sales, he will have an idea of the shape of the aggregate function. If he has some history to refer to, products which have been tried, shifts in formula, and their sales results, he will know more. Finally, the much deplored techniques of the pollster may help him in guessing the shape of the curve. In a world where only one man would be permitted to introduce new products, the data would probably be good enough to make success easy. In the real world where the manufacturer must not only guess the function but also outguess his competitors, it is extremely hard to make a profit.

Still, the task is not hopeless, and the existence of innumerable successful businesses demonstrate that a great many people have the necessary abilities to carry out the requisite calculations.

We now have a simple preference mechanism from which the preference schedule of conventional economic literature can be easily deduced. It also permits us to explain how people, with a set of basic drives that apparently evolved in the Stone Ages, can have tastes for innumerable things which did not exist then, and it accounts for the obvious fact that we have existing preferences for nonexisting things. Note, however, that the relative preference line drawn in Figure I is not a picture of the individual's underlying wants. It is instead a picture of the result of these underlying drives on his relative tastes for a continuously varying product. If the relationship between these underlying wants and the physical objects available in the real world is complex, as seems likely, then it is perfectly possible that the individual will make mistakes in these computations. Trial and error would rapidly eliminate such mistakes, but its possibility means that we can talk about the individual being mistaken about his preferences without absurdity.

It is something very much akin to this sort of "mistake" which was the main purpose for which Lancaster produced a very similar model.[11] His model, which is very close to a mapping of the one described above,[12] differs in two respects. Lancaster more or less explicitly uses characteristics which are real world variables rather than the rather metaphysical internal "wants" which I use. In addition, his model is used almost exclusively to discuss the purchase of different assortments of goods. His main theme is that consumers may, through experience and instruction, improve the efficiency with which they "consume." The point is important, and Lancaster's mathematics are most elegant,[13] but our model will be used to deal with other problems.

Note that the problem does not necessarily involve errors about the real situation. It is possible that the individual is simply misinformed and thinks that brand B is actually to the right of Y on Figure 1 and hence never tries it. Suppose instead that the individual knows the physical composition of the brands perfectly, but does not know his own preference curve. He has previously tried W, X, Y, and Z and, making a mistake in the computations connecting his wants, the experimental data he has, and the real world, thinks that the highpoint on his relative preference curve lies a little to the left of W instead of its right. He might, therefore, be initially more attracted by brand A than by brand B. No doubt he would quickly correct his error after trying A, but there would be at least a short period of time in which he was making a choice because he was in error as to his own preferences.

Errors of this sort would be most likely when the individual was choosing among unfamiliar alternatives, when the choice seemed to him unimportant so that he gave little thought to it, and when for some reason he was hurried. The experiments which have shown intransitive preferences normally involve all three of these conditions. The possibility that some of the apparent intransitivity they show may arise from mistakes of this sort, therefore, is a real one. Normally it would require a rather complicated set of errors to produce intransitivity, but since the phenomenon turns up only rarely, this does not rule out our explanation of the experimental results.

Our simple two-dimensional diagram may seem a very radical simplification of the real world, and indeed it is. Products and political platforms vary in many respects, not just one. Further, as we shall see, the one dimensional model may be very misleading in certain cases. Nevertheless, we shall return to it again and again since its very simplicity makes it an extremely useful intellectual tool. This means, of necessity, that each time we use it, we will have to carefully check to see that we have not been misled by oversimplification.

To proceed to a more general model, in Figure II, a product which varies in two particulars is shown, with the amount of each variable shown on one of the axes. For any individual there would be some combination of the two variables which would be optimal, such as the point marked. The quantities on the axes must, of course, be chosen so that all spaces in the figure are physically possible. The amount of agricultural subsidy could not be put on one axis and the expense on the other. If it is impossible to meet this requirement in some given case, then the fact that certain parts of the area are impossible should be indicated. The relative levels of preference for points other than the optimum can be shown by indifference curves, as on the figure. It seems reasonable to assume that the indifference curves would normally be concave toward the point of optimality as I have drawn them.[14] If this is so, then if we held one factor constant and varied the other, we would obtain a curve roughly similar to that in Figure I. This is the reason why I feel that the assumption of single peakedness is suitable for the single dimensional case.

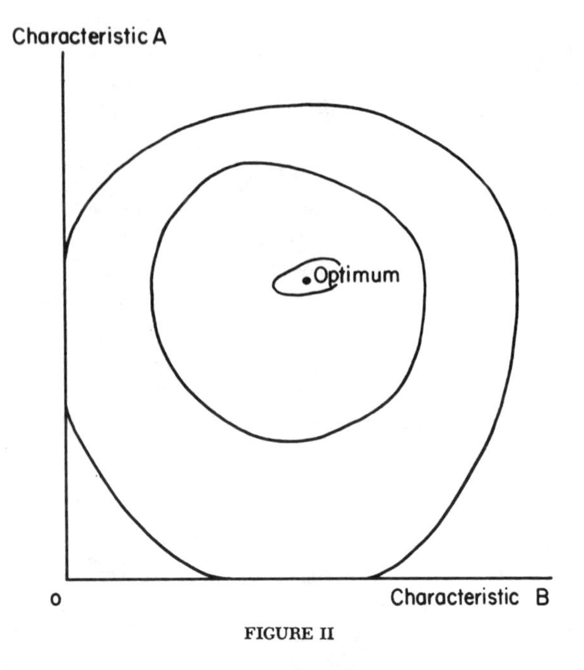

FIGURE II

All of the statements I have made about Figure I can be made about Figure II except the remarks about the simplicity of aggregating the preferences of several people. However, a discussion of this problem must be deferred until we have a better kit of intellectual tools. For the moment I can only say that I do not think that the problem is very difficult. Here also it would be possible for the manufacturer to guess what product would be preferred by the person whose optimum is as shown. Guesses would also be possible as to what new products or political programs would "sell." Again, the conventional ordinal preference schedule is deducible from our more elaborate system. It is obviously possible to extend the system from two variables to any desired number by the simple expedient of erecting further axes at right angles to these and locating optima in the resulting hyperspaces. Hypersurfaces would substitute for our indifference curves. Thus, our model can be fitted to any problem of choice, and can, therefore, serve as a base for our remaining analysis.

One problem remains. We have presented a model which provides "preferences" for new or potential products, but the model is essentially unchanging. It does not explain the use of resources to change tastes. A good deal of the advertising engaged in by both businessmen and politicians, of course, is simply informative, and does not involve any effort to change preferences. Before people can choose a given alternative, they must know that it exists. Thus, a new or improved product will normally require a good deal of "promotion." Further, the bone barrier to new information can be quite thick. Recently I had to change the prescription for my glasses. When I bought my new glasses I asked for prescription dark glasses since the clip-on dark glasses were, I thought, heavy and uncomfortable. The salesman, in almost an excess of honesty, informed me that this was not so, that very light plastic ones had recently come on the market. Although his statement was perfectly plain, I misunderstood it as referring to the older type and ordered the prescription dark glasses. About a month later, as I was trying to change from my regular glasses to the prescription ones while carrying three books, I suddenly realized that I might have misunderstood him. A visit to a drugstore convinced me that I had, and I no longer use my most expensive, prescription dark glasses.

Clearly, in this case, more than a simple statement was necessary to really inform me of the new alternative. Presumably this sort of thing is quite common, and actually getting people to consider new products or variations may be quite difficult. The rather extensive efforts to persuade housewives that they should try some product would seem to be examples of this sort of effort to inform. Free samples, reduced-price introductory offers, premiums, and very heavy advertising for new products are all efforts to get information through. Seldom are they intended to actually switch preferences. It is unfortunate but true that most of us are so conservative that it takes a great deal of persuasion to get us to try something new even when it is, by our own preference system, much superior to our established selection.

Another objective of advertising is actually to change the nature of the product. Eisenhower was particularly skilled at this sort of thing. A collection of ordinary measures would be named an "Eisenhower program," and would be accepted by the voters as somehow different from the same collection without the aura of the most successful public personality of our time. There seems no doubt that voter satisfaction was greater than it would have been, had the same measures been adopted without the public relations buildup. Commercial advertising can also have similar effects. Bufferin apparently does cure headaches somewhat better than it would if people were not told frequently on TV how good it is.

Similarly, some people get positive pleasure out of consuming something which is endorsed by the Beatles. In all of these, the physical nature of the product is unchanged, but its psychological nature is altered. Since we are interested in the satisfaction given, and that is psychological, the unchanged physical nature is of little importance. It is probable, however, that this effect of advertising is relatively minor. Most of the effort put into advertising, whether the advertising takes the form of a TV commercial or the President's State of the Union message, is aimed at informing the potential customer of an alternative or at persuading him.

This brings us to the last motive of advertising, a motive which has been both overestimated and badly misunderstood by a number of modern critics of advertising.[15] We may try to change people's tastes. So far we have accepted individual tastes as given, and have not asked how they were formed. When we turn to this question, we find ourselves back at the classical "nature or nurture" problem. The best opinion on this matter is that the mind of man is partly the result of hereditary genes, which need not be entirely the same for every man, and partly the result of his environment. This would apply to his preferences too, and it should, therefore, be possible to alter preferences by appropriate environmental changes. The inherited component, of course, puts limits on the variations which can be accomplished by education and propaganda. Still the limits are wide. Anyone who has spent much time with people of a radically different culture knows how much human desires are affected by the way men are educated.

In general, major molding of preferences is an authoritarian process imposed upon the individual. The education of the child, both formally in school and, more importantly, informally at home and at play, involves a great deal of indoctrination in the preferences thought "natural" in his culture. He will, for example, normally end up with a positive desire for a home life which is characteristic of the culture in which he has been brought up. Similarly, we make efforts, perhaps not too effective efforts, to "improve the taste" of children. In democratic cultures there is little of this authoritarian indoctrination of adults. Prisoners may be subjected to efforts to reform them and soldiers are normally given a good deal of indoctrination, but these are clearly exceptional situations.[16]

In nondemocratic cultures, and it must not be forgotten that these are overwhelmingly the dominant type in history, the state normally uses its control over all channels of communication to try to mold the preferences of adults along approved lines. Modern China and Russia will serve as models. These two examples also show clearly the limits which "nature" puts on the molding of preferences by education and propaganda. Given almost ideal conditions the "engineers of souls" have produced people who are a great disappointment to the governments. Clearly, the inherited desires in human beings put great limitations on the plasticity of human character. Indoctrination can do much, but it is not an unlimited tool. Further, the easiest time to influence a person is when he is a small child. We do not need to be an enthusiast for diaperology to realize that it is much easier to instill a taste for something in a child than to change that taste when the child has become an adult. The really major effect of the cultural environment takes place in childhood. With the child, the only limitation on what can be done by character molding is his genetic inheritance. With the adult, any effort to change preferences is limited by both his genes and his cultural indoctrination as a child.

Serious efforts to change people's preferences in a free society are apt to be successful only with that part of society which is not free. The child legally compelled to go to school and be indoctrinated in what the majority of the community thinks a child should learn is the classical case. The free adult is free, among other things, to refrain from listening to his moral or commercial mentors. Repetition and authoritarian pronouncements, so important for major preference shifting, are unlikely to attract an audience which is free to go elsewhere. The politician or the advertisers of a commercial product must always remember that the audience can shift to another channel with a simple twist of the knob. They must attract people in competition with other entertainments, and this puts very strict limitations on the methods which can be used to shift preferences. Nevertheless, it seems likely that at least some shifts in tastes are effected by political and commercial advertisers.

Suppose that a businessman or politician is considering introducing a new product or political proposal. For simplicity let us assume that he is attempting to attract only one customer-voter, Mr. A. The entrepreneur can select something which already stands higher in Mr. A's preference system than any presently available alternative, he can select something which will require a small change in Mr. A's tastes, or he can select something which will require a major change. Clearly the advertising costs of selling the new item will be lowest in the first case and highest in the last.

The entrepreneur would always choose the first alternative if he could, so a choice of the second must be put down to some barrier, probably technological, which makes it impossible or very expensive to produce the product which A would readily choose. The producer feels, perhaps, that the combined cost of producing the second commodity and "selling" it to Mr. A is less than the cost of producing and selling the first product even though the sales costs will be much less. Similarly, the entrepreneur would have a motive to choose the alternative which required the smallest change in Mr. A's preferences. In general, new products which require no shifts in preferences are more likely to be chosen by entrepreneurs than those that do require such shifts. If changing preferences is worthwhile, then small changes will be preferred to large, and changes in areas where preferences are not strongly held, to changes in strongly held preferences.

In a strictly technical sense, this process of attempting to change preferences only when such a change seems fairly easy will rationalize the individual's preference schedule. Where there is some tension between parts of the individual's preference ordering, changes which reduce that tension would normally be fairly easy, and, hence, we should expect entrepreneurs to prefer such changes. For example, we like both to feel that we are doing our duty and to receive benefits. President Kennedy played on these two, normally conflicting, desires by instructing us to "ask not what your country can do for you, but what you can do for your country" and providing a long list of things we could do for our country. Each of the items on the list was the acceptance of a benefit from the government.

One likely cause of internal tension in the preference ordering of an individual arises from the "nature-nurture" problem. An individual's preference system is the resultant both of an inherited component and an environmental component. The inherited component is not subject to change by advertising or propaganda, but the environmental component is. It would, therefore, be cheaper to try to move the preference schedule in the direction of the inherited component. There is some truth in the view that advertisers are vulgarizing our culture. They are engaged, at least to some extent, in moving us closer to "natural man" and weakening the educationally imposed layer of culture.

The picture of the preference structure of the individual sketched in this chapter retains the relative emptiness of content which characterizes the preference schedule of traditional economics. It is, therefore, equally hard to refute by psychological means. It is, in fact, simply an underlying structure from which the preference schedule may be deduced. The system is not only simple and obvious, it is also testable. We need only obtain a collection of objects which are identical in all respects but one, and then plot the relative preferences expressed by some individual for them with the varying characteristic as one axis of the figure. The experiment is simple, and the outcome so obvious that I doubt that it will be performed.

CHAPTER II

Pencil Exercises

For the economist the Edgeworth box is a common intellectual tool, but since it may be unfamiliar to the political scientist, some explanation is in order.[1] Suppose we have one hundred oranges and one hundred apples and two persons, A and B. The apples and oranges are distributed between A and B in some way. The situation may be shown diagramatically in the Edgeworth box of Figure III.

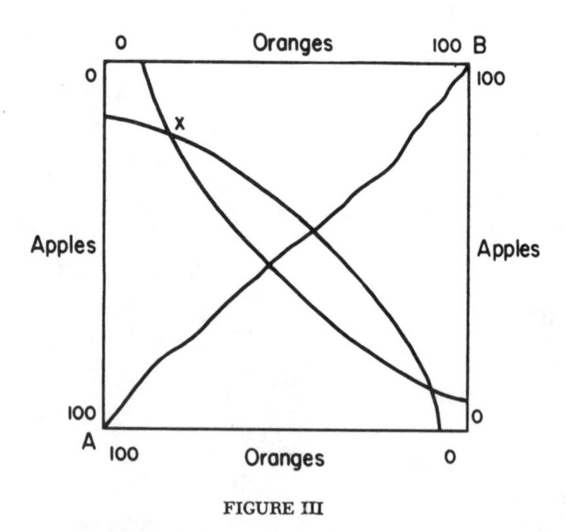

FIGURE III

The left and bottom sides of the box show how many apples and oranges, respectively, A has, and similarly for the right and top sides for B. Thus, point x simultaneously represents about ninety oranges and fifteen apples for A and about ten oranges and eighty-five apples for B. In the same way, any point in the box will partition the apples and oranges between A and B.

Given that we start with some such point as x, the other points in the box can be divided into three groups from the standpoint of A. There will be those that A prefers to x, those that he thinks are worse than x, and those which he thinks are just as good as x. The line drawn through x concave to the bottom left of the box represents the partition between the points which A prefers to x, which lie below and to the left, and the points which are to him inferior to x, which lie above and to the right. Those points which he thinks of as indifferent to x, if there are any such, will lie along the line, which is called an indifference curve.

Note, however, that apples and oranges are essentially indivisible units. It is unlikely that one of the parties will hold 90.27 apples. If we consider the apples and oranges as being available only in wholes, then the surface of the box is not entirely available. Only 10,000 points on the surface exist. Under the circumstances, the indifference curve can still partition the space, but large parts of it may not represent points that are indifferent, but only a partitioning line between points that are preferred and those that are inferior. The exact shape of the curve in such areas will be arbitrary and curves in slightly different locations could be drawn, but this fact is of little practical importance. Needless to say, continuous variables are possible, and in such cases the line simply consists of all points which are indifferent to x in the view of one party.

The line passing through x concave to the upper right corner is an indifference curve for B similar to the indifference curve for A. The lozenge-shaped area bounded by the two indifference curves is, from the standpoint of both A and B, superior to point x. It is, therefore, reasonable to suppose that the two parties, if left to themselves, will find some way to rearrange the partitioning of the apples and oranges among themselves so as to move from x into this area. The exact point to which they will proceed will depend upon the mechanism they use to move and their respective skills in bargaining, but something can be said about their ultimate destination. If we draw indifference curves through any given point in the Edgeworth box, we will always find a lozenge of superior points unless the two curves are tangent to each other. Thus, unless the curves are tangent, some change which is an improvement from the standpoint of both parties is possible. Such changes will eventually lead to a point where the two curves are tangent.[2] The locus of all points where the curves are tangent will be a line, called the contract locus, and this line is normally drawn running between the two corners as in the diagram.

This way of drawing the contract locus, and the exact shape of the indifference curves shown in Figure III, however, is not quite suitable for our present purposes. It is quite sensible for the economist to draw it this way because he is normally only interested in the center of the diagram and the results he gets are correct for normal economic operations, but we are going to use this essentially economic tool for noneconomic purposes, and we must make some changes. It is not normally true that individuals are completely indifferent to the welfare of others. A would really rather not see B starve to death, hence his point of optimality is not the lower left corner, but some other point such as A in Figure IV.[3] The indifference curves, then, would take the roughly circular shape shown and the contract locus would connect the two optima of A and B. The indifference curves would have points of tangency to the right of B, indicated by the dotted line, but this would have no relevance as long as the two parties were free to

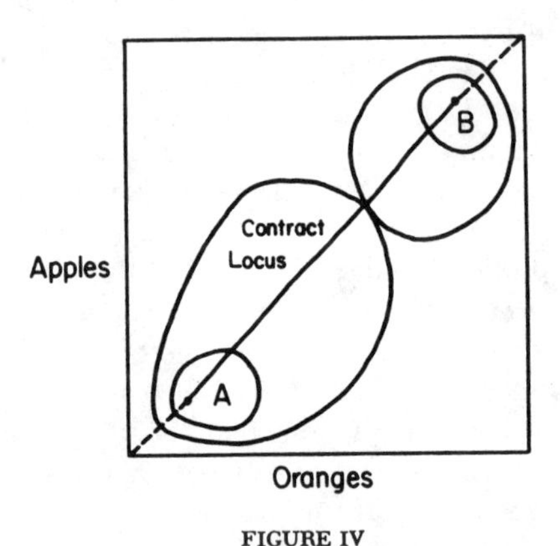

FIGURE IV

adjust their affairs themselves since movement from this line to B would be mutually beneficial—similarly for the points of tangency between A and the "A corner." Later we will find some use for these dotted extensions of the contract locus in cases where more than two parties are involved.

The contract locus between A and B is the Pareto optimal area, which simply means that if you start at some point off the line, it is always possible to reach the line in such a way as to either benefit both A and B or, at least, to benefit one of them without injuring the other. Once the Pareto optimal region is reached, however, any movement will certainly injure one or the other. Normally economists discuss the process of reaching the Paretian area, but do not discuss possible movements within that zone. Politically, with various voting rules in use, this is not enough, and we must consider movement between points which are Pareto optima. Before beginning this discussion, however, a digression on the geometric properties of the type of issue space which we will be using is desirable. This discussion will be relatively simple and nontechnical. Those interested in a more rigorous approach should read Ragnar Frisch, "On Welfare Theory and Pareto Regions."[4]

For our less rigorous analysis, let us consider a country club contemplating a fireworks display for the Fourth of July. It is proposed that this exhibition be financed by a per capita assessment of the members. We can display the amounts of money to be spent for this purpose on a straight line as in Figure V, and let us initially assume that the president appoints a two-man committee to decide how much shall be assessed and spent. The optimum amounts to each of the two individuals are shown at A and B. Let

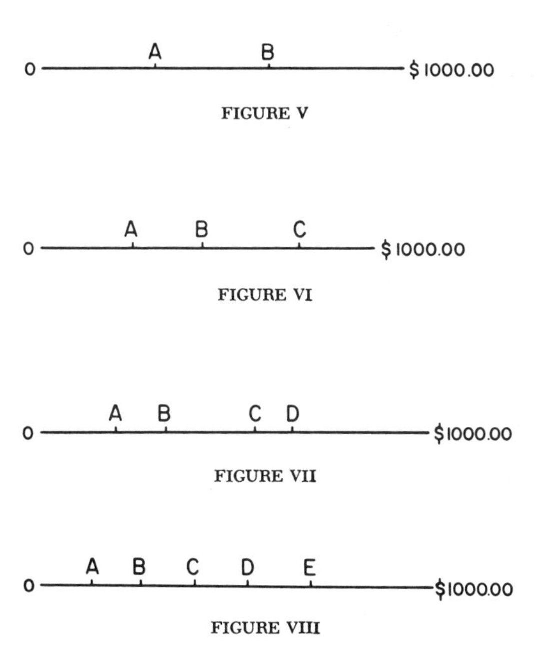

FIGURE V

FIGURE VI

FIGURE VII

FIGURE VIII

us further assume, as seems reasonable, that each of the two members of the committee get less and less satisfied as the amount to be spent becomes more distant from his optimum. Under the circumstances the Pareto optimum area is simply the segment of the line between A and B, since any movement along this line will reduce the satisfaction of one or the other. Movement to this segment from those parts of the continuum which are outside it, on the other hand, may benefit both of them, and there will always be at least one point on the Pareto optimality segment which will be superior to any point outside it.

If the committee is composed of three members, we can display them similarly as in Figure VI. Here the Pareto optimality segment extends from A to C. A four-man and a five-man committee are represented on Figures VII and VIII, respectively. The Pareto regions extend from A to D on the first and from A to E on the second.

Let us suppose that it is decided to also serve a meal to all members who attend the fireworks display, the cost to be included in the per capita assessment. The committee(s), then, will have to decide upon two issues at once, how much to spend on fireworks, and how expensive a meal. We can display the possibilities by erecting a second axis on our diagram to represent the various amounts of money which might be spent on the dinner. Figure IX shows this situation for a two-man committee, and Figure X, for a three-man committee.

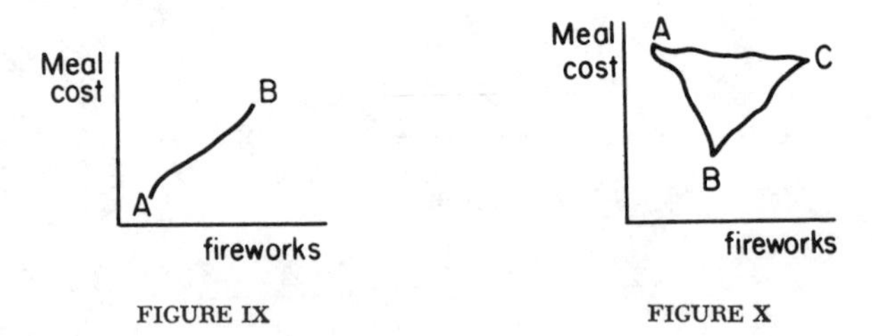

FIGURE IX FIGURE X

If the optima of A and B are as shown on Figure IX, then the contract locus connecting them is the Pareto optimal region. For three people, there are three contract loci, as shown on Figure X. In this case, however, the Pareto optimal region includes not only the three contract loci but also the area bounded by them. At any point in this area, any move is bound to injure at least one of the three parties. The situation for four-man and five-man committees are shown on Figures XI and XII.[5] We connect all the optima with contract loci, and once again the area surrounded by the contract loci is the Pareto optimal region.

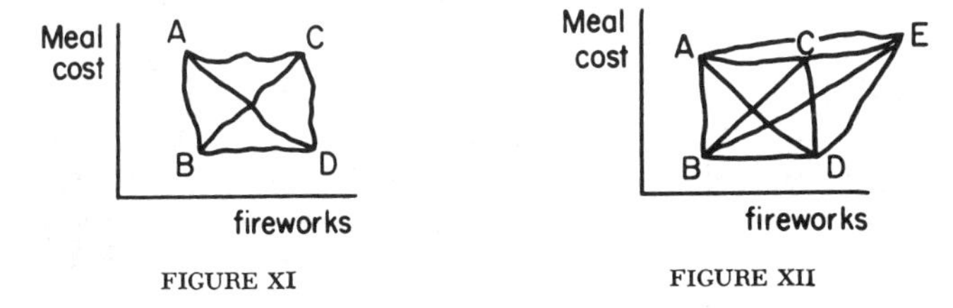

FIGURE XI

FIGURE XII

When I have presented this chapter as a lecture, I turn at this point to a three-dimensional tic-tac-toe board in order to show the arrangement of the Pareto optimal space in three-dimensional space. This is, of course, impossible in a book, so I must ask the reader to use his ability to visualize objects in space to continue along the next step of our reasoning. Suppose, then, that a third variable, say the amount to be spent on drinks, is added. We represent this by erecting a third axis at right angles to the first two and examine the positions of the members of the committee in this new three-dimensional space. For the two-man committee, the roughly spherical indifference surfaces around their optima will still be tangent at a contract locus, which is simply a line connecting them. For three men, the triangular Pareto optimal surface will also still exist although it may not lie entirely in two dimensions. For the four, on the other hand, the figure changes, and instead of a rough square we will have the shape of a rough tetrahedron, with each of the four surfaces of the figure connecting three optima. The Pareto optimal region of the five-man committee will have one of several shapes but will still have contract loci and surfaces which connect three optima running through the Pareto optimal region.

Certain regularities should, by now, be obvious. If the number of dimensions is less than the number of individuals minus one, then the figure described by the Pareto region will be "compressed" with contract loci and surfaces connecting optima running through the Pareto optimality zone. When the number of dimensions is equal to the number of people minus one, the figure formed by the Pareto region assumes a characteristic form, which it will hold also if the number of dimensions is equal to, or greater than, the number of people choosing. This form is a line in the case of two people, a triangle in the case of three, a tetrahedron in the case of four, and higher order figures in higher order spaces for committees larger than four.

Returning to Figure V, it will be observed that about one-third of the possible decisions lie within the Pareto optimality area. In Figure VII, which also shows a committee of two, but with two issues, the total size of the Pareto space is greater, but as a percentage of the whole issue space, it has shrunk to insignificance.[6] If we continue to a three-dimensional continuum, the portion of the total issue space occupied by the Pareto region shrinks still farther, and similar shrinkage would be expected as the number of dimensions grows.

If we consider the three-man committee dealing with a single variable as shown on Figure VI, and with two variables as shown on Figure X, it is clear that the percentage of the total area of possibilities which make up the Pareto optimal area has shrunk, but not very greatly, with the addition of the second dimension. In a three-dimensional issue space, on the other hand, the percentage of the total space which would be included in the Pareto region shrinks very sharply. The same line of reasoning can be followed with the larger committees. As the number of issue dimensions increases, the Pareto optimal portion of the total area shrinks relatively slowly until the number of issues is one less than the number of members of the choosing group, and then shrinks very rapidly thereafter. Assuming a given number, n, of voters and varying dimensions, the percentage of the total space which would be Pareto optimal can be graphed as in Figure XIII.

Figure XIV shows the situation if the number of issues is held constant and the number of voters is varied. Absolute magnitudes are shown in Figures XV and XVI.

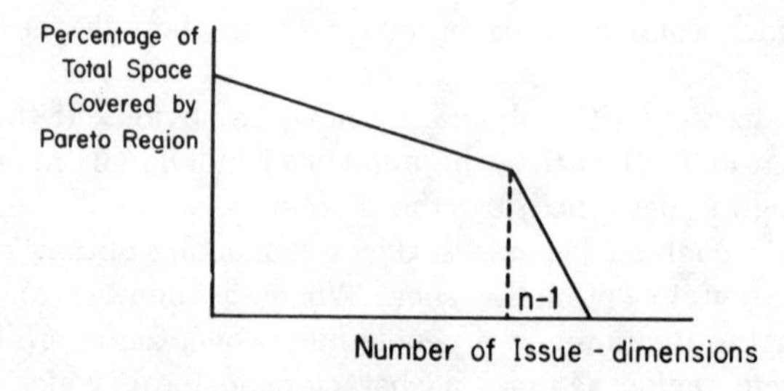

FIGURE XIII

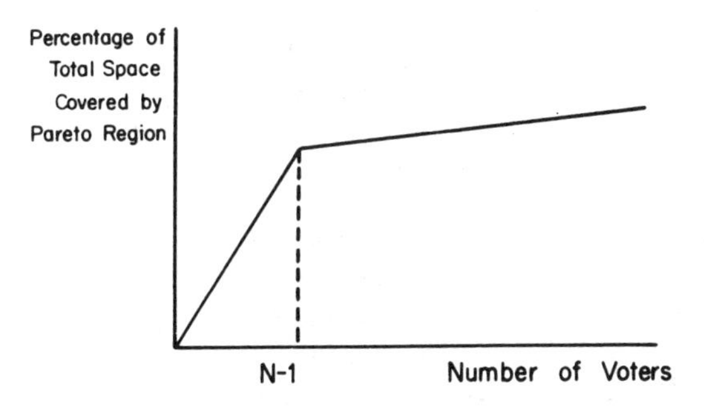

FIGURE XIV

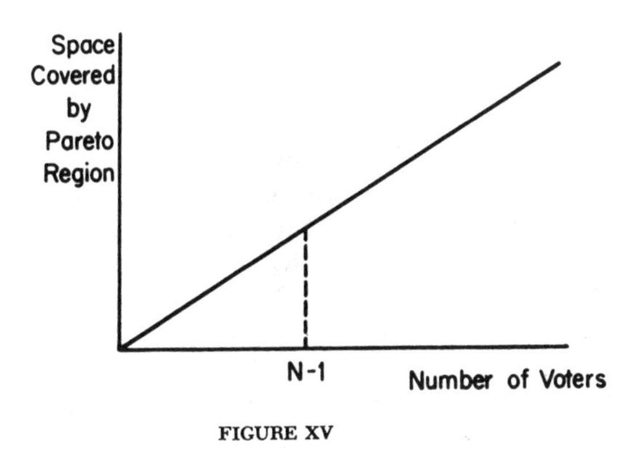

FIGURE XV

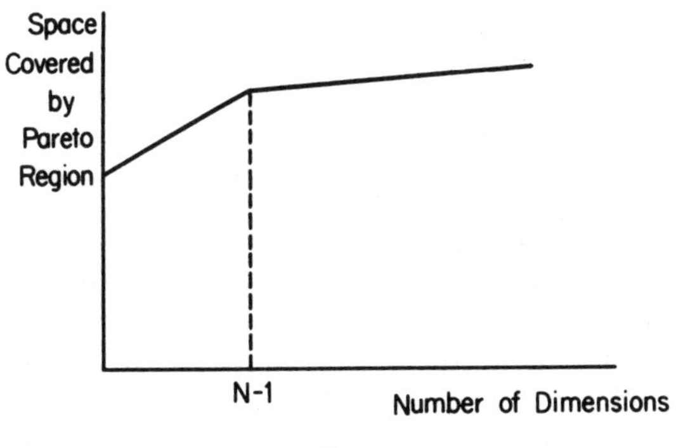

FIGURE XVI

From these two figures it is clear that where the number of issue-dimensions substantially exceeds the number of choosing parties, the Pareto region is relatively very small. Under these circumstances the goal of the economist, reaching the Pareto frontier, seems very sensible. Fortunately, in most economic matters this situation obtains. The ideal economic transaction involves only two persons, a buyer and a seller, and will involve quite a number of variables concerned with exactly what is to be exchanged.

Where the number of choosers greatly exceeds the number of issue dimensions, however, the Pareto region may include a very large part of the issues space, indeed, almost all of it. In such circumstances merely reaching the Pareto frontier may seem a very modest objective. We feel, somehow, that certain areas upon the Paretian possibility frontier are to be preferred to others. The traditional Paretian analysis, however, is of no immediate utility in determining where such preferred areas would lie. In order to discuss this issue, we must turn to a variant on the Paretian system introduced by Dr. James Buchanan and myself in *The Calculus of Consent.*[7]

Let us consider, not an individual choice problem, but a rule to be used in reaching decisions in a long series of such problems, with the exact details of most of the specific decisions unknown. If we feel that we cannot make interpersonal comparisons of utility, then it is very hard to fault the Wicksell version of the Paretian criterion, unanimous agreement for any given decision. If we are thinking of a decision rule for a multitude of choices, probabilistic considerations come in. We would be perfectly willing to choose a rule which might sometimes injure us, if we are convinced that it will sometimes benefit us and the present discounted value of the benefits exceed the injury. The optimal rule would be the one in which the excess of the discounted benefits over the discounted injuries was greatest.[8] How we choose such a rule has been discussed in *The Calculus of Consent,* and will not be repeated here. Suffice it to say that unanimous consent would seldom, if ever, be selected as the optimal rule. Normally some kind of voting rule in which some specified number of choosers are permitted to impose their will upon the remainder will be chosen.

Such nonunanimous voting rules are not very desirable in cases where the number of issue dimensions exceed the number of voters minus one. In such cases, the use of almost any voting rule will simply lead to some point on the outer boundary of the Pareto region.[9] A demonstration of this will be presented below for three choosers in a two-dimensional issue space, and it is intuitively obvious that the same rule would apply for the higher dimensions. This does not seem any improvement over the unanimity requirements; in fact, it might be regarded as worse since the unanimity criterion would occasionally, when the starting point happened to be inside the Paretian zone, lead to a solution in the interior rather than the boundaries of the Paretian region.

If the number of choosers greatly exceeds the number of issue dimensions, on the other hand, the interior of the Paretian region will be thickly cut by contract loci and hyperplanes connecting the various optima. In this situation almost any voting rule will result in at least some movement toward the center of the zone. Movement from the boundary of the Paretian region toward the center will always benefit more people than it injures[10] and hence there is some presumption that in choosing a rule for a large number of future decisions, we would regard this as advantageous. The tendency of any voting rule to bring the ultimate result nearer the center of the Paretian region has the result that people whose tastes approximate the average of their society are likely to have their optima much more closely approximated than nonconformists.

The ultimate outcome of this line of reasoning, then, is that in situations where the number of choosers greatly exceeds the number of issue dimensions, voting rules are apt to be chosen. This, then, is the proper domain of political choice. We have previously seen that where the converse situation obtains, where the number of issue dimensions greatly exceeds the number of choosers, the unanimity criterion is apt to work well. How about the area in between, where the two quantities are of the same order of magnitude? Unfortunately, neither method of reaching a decision works very well here. This is not only unfortunate for us as inhabitants of the real world, it is also unfortunate for us as students of decision processes. Simple two-dimensional issue spaces with small numbers of choosers, say from two to seven shown on them, are easy to construct and study. But this is precisely the area where the two systems are of limited value. Thus though these studies are fascinating in themselves and surely cast at least some light on the real world, they are less significant than they at first appear. Nevertheless, they are easy to do, and this may excuse work in this field. If the payoff is low, so is the cost.

Let us return to the Edgeworth box and consider two individuals, A and B, whose preferences are as indicated on Figure

XVII. If we suppose that the two parties begin at point X, it is not hard to predict that they will move to some point within the lozenge, and we can even feel fairly confident that they will end up on some part of the segment of the contract locus between the two indifference curves going through X.

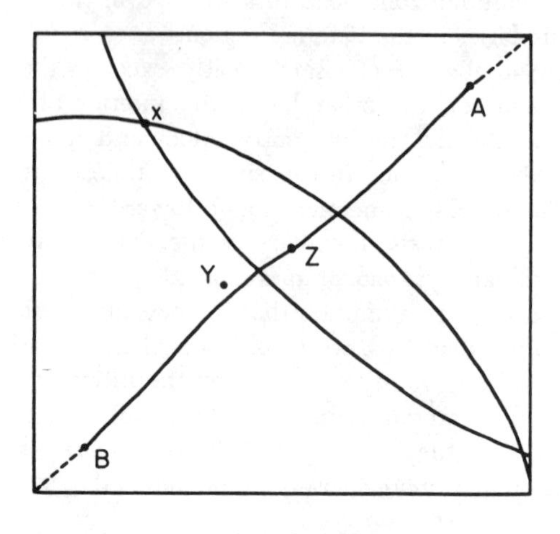

FIGURE XVII

Points X, Y, and Z illustrate a problem raised by Kenneth Arrow.[11] If we assume that A and B make up their mind on changes by majority voting,[12] then Y could not get a majority over X, nor Z over Y, but Z could get a majority over X. Arrow points out that this result is intransitive, even though the individual preference orderings are transitive. The intransitivity of the result is, of course, clear, but it doesn't seem particularly remarkable when shown in the Edgeworth box. Arrow puts his problem in a more general form, and simply presents a pair of individual preference orderings as follows:

A	B
Z	Y
X	Z
Y	X

The difference between the two systems is that Arrow assumes complete independence among the individual preference ordering, while our use of the Edgeworth box implies a special form of dependence. This form of dependence, in which it is simply assumed that there is some issue space in which each individual will have an optimum, and that his relative favor for points other than the optimum is a function of the distance from the optimum, will be acceptable by most economists as fairly descriptive of the real world. The interdependence comes from the fact that the optima are in the same space. In other words, it comes not from any identity of the individuals, but from the fact that they must choose from among the same set of alternatives and that these alternatives have certain physical characteristics. The general shape of the preference map of the individuals, roughly a mountain, is also assumed to be the same, but this assumption will not raise much contradiction.[13] This rather limited type of interdependence will be assumed for the remainder of this book except where some other assumption is specified.

If the situation shown on Figure XVII were either a political or an economic problem, the details of the procedure used would be the same. One of the parties would suggest movement from X to some other point. The other party would either accept or reject this proposal. If the proposal were rejected, other proposals would be made until one was accepted. In fact, even if the proposal to move were accepted, a further proposal might well be made. We shall generalize this procedure for our future analysis by assuming that the individuals, and they will be very numerous in some future problems, make proposals either in rotation or randomly, that these proposals are then decided by whatever decision process is under discussion, and that further proposals are then made until some sort of stopping place is reached.

Given preferences as shown in Figure XVII, if A proposed movement to Z, B might refuse. This would not be because he would be worse off at Z, but because he thought that by rejecting this alternative he might be able to persuade A to let him reach an even more favorable outcome, say some point between Z and the intersection of the contract locus and A's indifference curve through X. This would be an effort to bargain, and we will spend very little time on such matters. Generally, let us assume that individuals "vote" for alternatives they prefer and do not reject such alternatives in hopes of getting something even better. Our justification for this assumption is not that bargaining of this sort is little understood, although that is true, but that with many possible parties, bargaining of this sort is of little importance.[14] Since we are analyzing cases with few parties, where bargaining is important, only in order to extend our analysis to cases where the parties are numerous, we can afford to ignore, except in a few cases, this "small number" phenomenon.

Although we cannot, strictly speaking, use the Edgeworth box to represent a partition among three parties, we can use it to represent three, four, or any number of person's optima on any set of two variables.[15] If the two dimensions in Figure XVII, for example, represent apples and oranges and each point represents some partition of them between A and B, then C might have preferences on this point. He might, for example, be a radical egalitarian and feel that the apples and oranges should be evenly divided between A and B. If this were so, his optima would be at the exact center of the figure. If apples and oranges were distributed between A and B by some sort of collective decision process in which A, B, and C all voted, this sort of representation would be most useful. A great many distributional decisions, of course, are taken collectively; in fact, redistribution of wealth is one of the principal activities of the modern state.

Normally, however, these problems are discussed in terms of some sort of public good other than redistribution, and we shall adhere to this custom. On Figure XVIII the two axes represent ex-

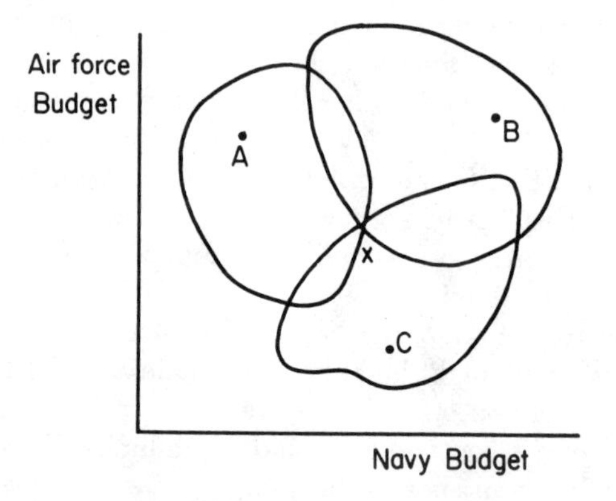

FIGURE XVIII

penditure on two different arms of our defense establishment, and the optimum and one indifference curve are shown for each of three individuals. All three indifference curves go through the point marked X on the figure, and it can be seen that there are sizable areas, shown by the lozenges, which would be preferred by two voters to X. This is, of course, Black's classically simple proof that there is no majority motion in this situation. It is reasonably obvious that without very strong restrictions on the shape of the indifference curves, no point on the surface will be able to get a majority over all other points. Thus, if majority voting were resorted to, there would be no ultimate end to the process; the possibility of proposing another motion which could get a majority against any chosen position would always exist.

Actually, however, if the motions to be voted upon were also made by three individuals, either in turn or randomly, no point like X would be likely to be included among them. In Figure XIX the contract loci have been plotted. If we were to start at X, then we could be certain that some set of points on each of the contract loci could get a majority over that. Y, for example, could get a majority over X. V and T could also get a majority over X, but we would normally assume that if A and C were in a coalition to get the maximum benefit, they would make a series of proposals which would enable them to reach their contract locus. They would also vote down B's proposals. Note, however, that the presence of B, in a sense, reduces the bargaining difficulties in reaching the contract locus. Either A or C could abandon his attempts to make a coalition with the other, and turn to B as a partner. This means that both A and C must avoid being seriously obstructive. The situation is, in this respect, quite different from the two-party problem.

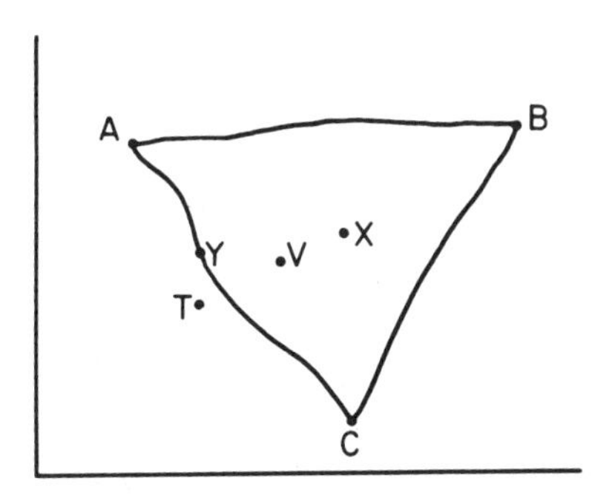

FIGURE XIX

Before turning to a discussion of the important problem of cycles, we should note that it is at least theoretically possible for the majority voting process to get outside the Pareto optimal area. T would be preferred by A and C to X, and hence they would vote for it if confronted with this choice. If we assume that some outside force is selecting the alternatives among which the parties vote, then this is clearly an open option to that outside authority. If, however, we assume that proposals for changes are made by the parties themselves, then there is no reason to expect that they will make proposals of this sort. Presumably, they would not know the exact location of the contract loci, and there would, therefore, be a process of trial and error, but they would have no motive to confine their choice set to points away from the contract locus, and points near the contract loci would normally win in the balloting. Thus, the likely result, granted that the people who are to vote also propose the alternatives, is a result on one of the contract loci.

Having decided that the result will be on the contract loci, however, we must still deal with the problem of the cyclical majority. The ordinal preferences of A, B, and C, on Figure XX among Y, W, and Z can be presented as follows:

A	B	C
Y	W	Z
Z	Y	W
W	Z	Y

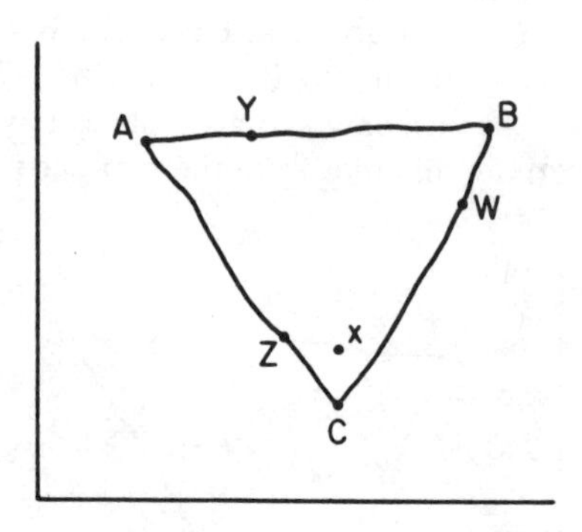

FIGURE XX

In this situation, W can get a two to one majority over Y; Z can get a two to one majority over W; and Y can get a two to one majority over Z. Thus, no one of this set of three motions can get a majority over any other. Further, every point on the three contract loci is part of a number of such cycles.[16] Thus, no definite point can be specified which would get a majority over all others. It is at least theoretically possible for the voting to get into an endless process of choosing Y over Z, Z over W, and W over Y.

It is also possible that some points off the contract loci might get involved, but this is not likely. X could get a two to one majority over W, but it is unlikely that either A or C would propose X as opposed to some point like Z which is on the contract locus. If they are reasonably sensible, they would simply never propose an inferior motion, and if they did make such a mistake, they would shortly correct it by proposing another motion taking them to the contract locus.

Cycling endlessly on the contract locus is not, however, a very attractive prospect. Is there any likelihood that the process would ever reach an end? Surely the three individuals would, with time, get tired of the cycle, and this might lead them to simply stop it. For example, if the result of previous voting were Z and a proposal to move to Y were made, A might decide that he would rather stop at Z than continue in an endless cycle. An agreement might even be entered into between two parties to shift to some mutually acceptable point and then to refuse to vote for any other, no matter how desirable the shift might appear from the standpoint of one of them.[17]

It seems possible, however, for an endless cycle to arise in cases such as the one under discussion. In practice, it is fairly certain that some procedure would be used to end the cycle, but, there being no rule for determining where the termination would come, this is little different from saying that the cycle would be stopped at a randomly selected point. Thus, it would appear that democracy is almost a random choice process. In the next chapter it will be demonstrated that this is not so if the number of voters much exceeds the number of issue dimensions, but it probably is true with small numbers of voters. This is a further reason for feeling that in the cases where the number of issue dimensions and choosing individuals is somewhat of the same order of magnitude, voting (and economic choice) works badly.

We can apply the same sort of analysis to larger numbers of voters. Four voters could be as shown in Figures XXI and XXII. Four voters on a two-issue dimension graph have many desirable

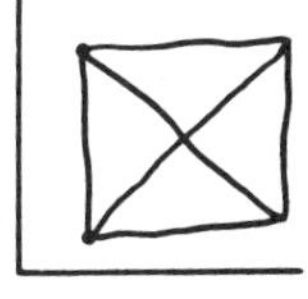

FIGURE XXI

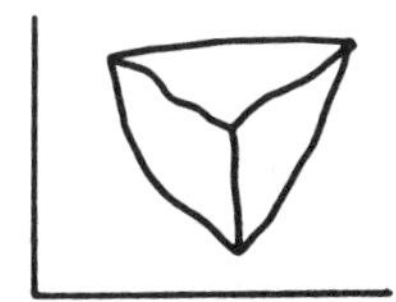

FIGURE XXII

characteristics. It is almost impossible to work out cycles. Under almost any voting procedure the group choice will rapidly move in to the intersection of the two diagonals if the voters are arranged as in XXI, and to either the central voter or some point on the extensions of the contract loci near his position if the voters are arranged as in XXII. Unfortunately, all of these fine features are characteristics only of the two-dimensional situation. In three dimensions, the figure opens out to a tetrapod, and loses all of these advantages. Since there is no way of restricting voting to two-issue dimensions, this reduces the practical importance of the theoretical neatness of the four-voter two-issue case to almost zero.

Five voters can be arranged in several basic arrangements, the most important of which are shown in Figures XXIII, XXIV, and XXV. A good deal of innocent amusement can be derived from the examination of these cases, and determining likely outcomes of

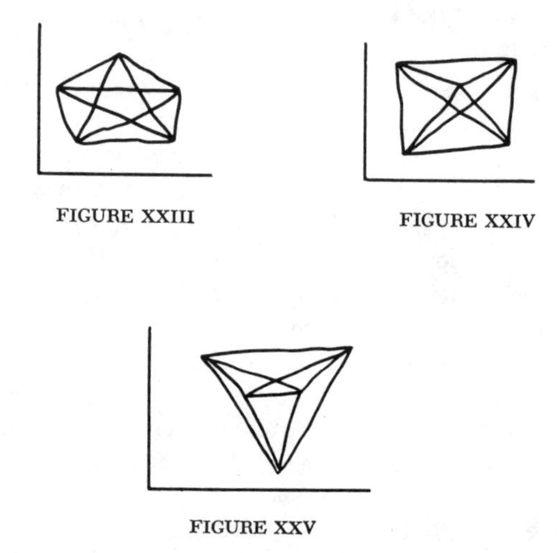

FIGURE XXIII

FIGURE XXIV

FIGURE XXV

voting is an excellent drill. In Figures XXVI and XXVII the areas which would get majorities against two chosen points are shaded. Areas which can get a four-fifth majority are crosshatched.[18] It will be noted that there are no positions which can get a majority over all others and that there are numerous positions outside the Pareto optimum area which can get majorities over areas within it. A somewhat less obvious characteristic is also clearly shown. Among the contract loci there are some which only fail to bisect the optima because there are an odd number of optima. These loci

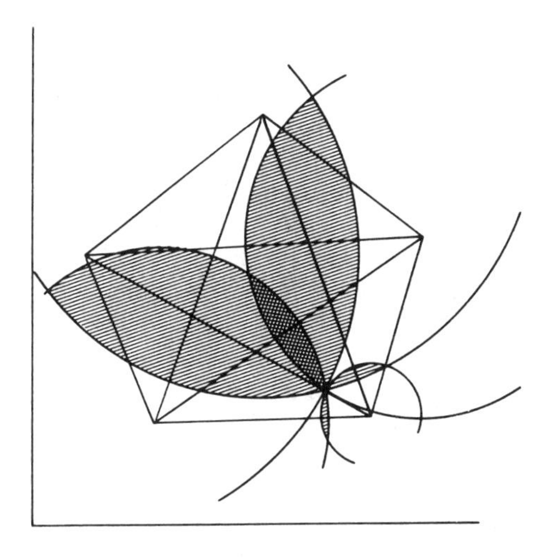

FIGURE XXVI

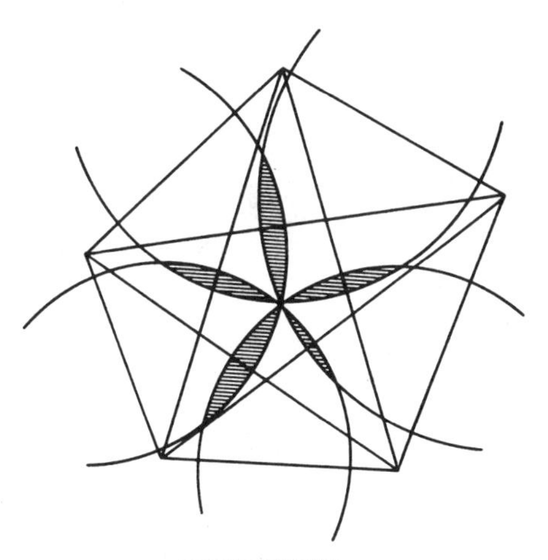

FIGURE XXVII

each have only one more point on one side than on the other. The areas which can get a majority over any given point are lozenges pointing at such lines and roughly bisected by them. Whether this fact is important or not, I cannot say. Personally I have found the type of exercise shown on Figures XXVI and XXVII both interesting and amusing. I cannot claim, however, that I have learned anything very important from it.

In the next two chapters we will shift to many voters, choosing in a two-dimensional issue space. This is more suitable for the voting process than the relatively small number of voters we have been examining here. Further, provided only that the number of voters remains much in excess of the number of issue dimensions, these results will be easily generalizable to more than two-issue dimensions. One most important phenomenon of politics, however, cannot be presented in two-dimensional issue space. This phenomenon, log-rolling, will have to be deferred for later treatment. This is most unfortunate, but the test of a model is its results, not its premises, and our model which rules out log-rolling and, hence, is highly unrealistic, will give quite good results. The reason for these good results will be explained in the latter part of the next chapter, but some indication may be given here. Briefly, log-rolling must be dealt with by a multidimensional generalization of the two-dimensional model. Thus, our simple two-dimensional system can be taken as an analogue of the higher dimensional models which fit reality with quite good precision.

CHAPTER III

The General Irrelevance of the General Impossibility Theorem

A phantom has stalked the classrooms and seminars of economics and political science for nearly fifteen years. This phantom, Arrow's General Impossibility Theorem, has been generally interpreted as proving that no sensible method of aggregating preferences exists.[1] The purpose of this essay is to exorcise the phantom, not by disproving the theorem in its strict mathematical form, but by showing that it is insubstantial. I shall show that when a rather simple and probable type of interdependence is assumed among the preference functions of the choosing individuals, the problem becomes trivial if the number of voters is large.[2] Since most cases which require aggregation of preferences involve large numbers of people, "Arrow problems" will seldom be of much importance.

In *Social Choice and Individual Values,*[3] Arrow included a chapter on "Similarity As the Basis of Social Welfare Judgments"[4] in which he discussed possible lines of research which might lead to a method of avoiding the implications of his proof. In this chapter, he pointed to Black's single-peaked preference curves as particularly promising.[5] The generalization of Black's single-peaked curves to more than one dimension will give the fundamental model upon which this article is based. It may be fairly said, therefore, that the present work follows the path indicated by Arrow. The development of single-peaked preferences for two dimensions was first undertaken by Newing and Black in *Committee Decisions with Complementary Valuation,*[6] which was published at about the same time as *Social Choice and Individual Values* and presumably not known to Arrow. Newing and Black, however, did not give much consideration to cases in which there were large numbers of voters. The model to be used here will involve many voters and will be used to examine the general impossibility theorem.

The proof of Arrow's theorem requires as one of its steps the cyclical majority or paradox of voting.[7] In addition to the mathematical reasons, the emphasis on the paradox is appropriate since the method of "aggregating preferences" which immediately occurs to the average citizen of a democracy is majority voting. This chapter is intended to demonstrate that majority voting will, indeed, always be subject to the paradox of voting, but that this is of very little importance. Majority voting will not produce a "perfect" answer, but the answer it does produce will not be significantly "worse" than if the paradox of voting did not exist. Any choice process involving large numbers of people will surely be subject to innumerable minor defects, with the result that the outcome, if considered in sufficient detail, will always deviate from Arrow's conditions. The deviation may, however, be so small that it makes no practical difference.

Most majority voting procedures have arrangements which bring the voting to an end before every tiny detail of the proposal has been subject to a vote. These restrictions (frequently informal rather than part of the rules of order) mean that when the voting is brought to a stop, minor changes probably remain in the result which a majority would approve if it were possible to bring them to a vote. Thus, the outcome will be, in Arrow's terms, imposed, but it will be very close to a perfect result. As an example, suppose a body of men is voting on the amount of money to be spent on something, with the range under consideration running from zero to $10,000,000. The preferences of these men are single-peaked. Majority voting will eventually lead to the selection of the optimum of the median voter as the outcome. If, however, the procedure is such that proposals to change the amount of money by $100 or less cannot be entertained,[8] then the outcome will normally not be at the optimum, but will be within $50.00 of it. This result does not meet Arrow's conditions, but there is no reason to be disturbed by this fact.

In order to demonstrate that the cyclical majority is equally unimportant in real world "preference aggregation," let us consider a group of voters deciding two matters, say appropriations for the Army and the Navy, by majority voting. In Figure XXVIII the vertical dimension is the appropriation for the Army and the horizontal, for the Navy. The individual voters each have an optimum combination and a preference mountain which has the usual charac-

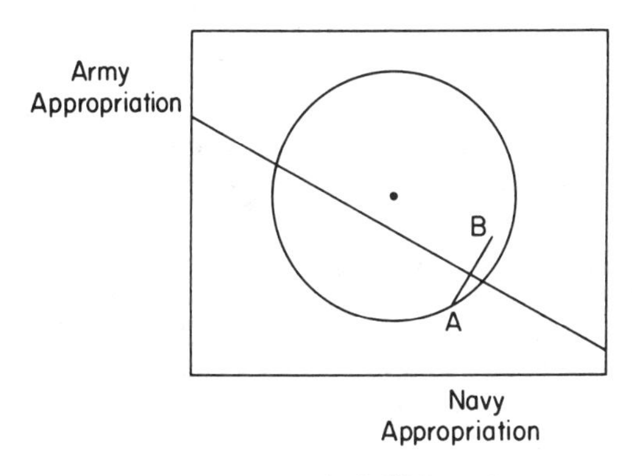

FIGURE XXVIII

teristics. For simplicity let us further assume that the voters' optima are evenly distributed over the space, and that their indifference curves are all perfect circles centering on their optima. The last two assumptions do not correspond with reality and will be eliminated at a later stage. Let us further assume that the number of voters is great enough so that the space can serve as a proxy for the voters. In other words, of two areas in the issue space of Figure XXVIII, the larger will contain the optima of more voters than will the smaller. This makes it possible to use simple Euclidean geometry as an analytical tool.

Suppose we wish to determine whether motion B on Figure XXVIII can defeat the status quo, represented by A, by a simple majority vote. Since we are assuming that all indifference curves are perfect circles around the individual's optimum, each voter will simply vote for the alternative which is closest to his optimum. If we connect A and B with a straight line and erect a perpendicular bisector on this line, then B will be closer to the optima of all and A will similarly be closest to all optima which lie on A's side of individuals whose optima lie on the same side of the bisector as B, the bisector. We can compare the votes for each alternative by simply noting the area of the rectangle on each side of the bisector. As a shorthand method, if the perpendicular bisector runs through the center of the rectangle, A and B will have an equal number of votes. If it does not, then the alternative on the same side of the perpendicular bisector as the center will win. The locus of all points which will tie with A is a circle around the center running through A, and A can beat any point outside the circle but will be beaten by any point inside. Clearly, no cycles are possible. The process will lead into the center eventually, since, of any pair of alternatives, the one closer to the center will always win.

This might be called the perfect geometrical model, in which the number of voters whose optima fall in a given area is exactly proportional to its area. Given that the voters are finite in number, small discontinuities would appear. Two areas that differ little in size might have the same number of voters; indeed, the smaller might even have more. Cycles are, therefore, possible, but they would become less and less important as the number of choosing individuals increases. In Figure XXIX we have a point, A, in our standard type of issue space, and I have drawn a circle around

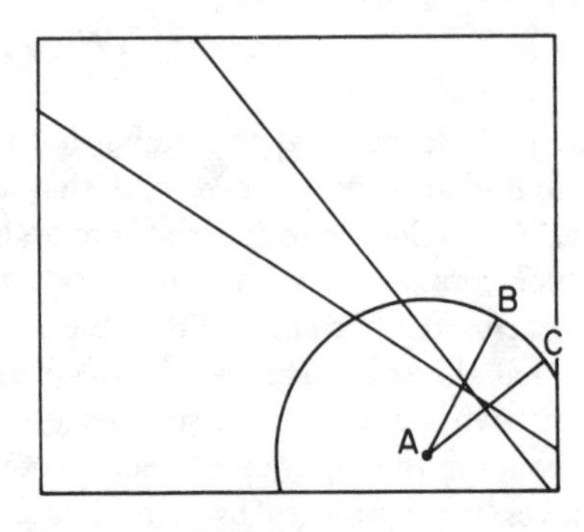

FIGURE XXIX

it. For convenience I shall assume 999,999 voters. Whether any given point on this circle can beat A depends upon how the perpendicular bisector of the line connecting it with A partitions the voters' optima. We can conceive ourselves as moving around the circle, trying each point on it against A. B would beat A, but C would not. With a finite number of voters, the changes between motions would be discontinuous. B, for example, might get 602,371 votes to A's 397,628. As we moved around the circle toward C, there would be a small space in which this vote would stay unchanged, then it would suddenly shift to 602,370 against 397,629, which would also persist for a short segment of the circle. Needless to say, the segment in which the vote did not change would be extremely small, but it would exist.

If we consider a point which gets only a bare majority over A, 500,000 votes to 499,999, and move along the circle toward B, there will be a finite gap during which the vote does not change, and then it will shift to 500,000 for A and 499,999 for the alternative.[9] Given this finite distance, however, it is sure that at least occasionally a point can be beaten by another point which is more distant from the center than it is. In other words, it will be possible for majority voting to move away from the center as well as to move toward it. This phenomenon makes cycles possible.[10]

Granting these discontinuities, however, we could still draw a line separating those points which could get a majority over any given point from those that could not. With our 999,999 voters, this line would no doubt appear to the naked eye to be the circle of Figure XXVIII. Examining it through a microscope, however, we should find that it was not exactly circular and that there would be small areas which could get a majority over the original point, but which lay farther from the center than that point. Note, however, that these areas would be very small. If our original point is far from the center (as is A in either Figure XXVIII or Figure XXIX), then the area which could get a majority over A but which lies fàrther from the center, would be tiny compared to the area which could get a majority and which lies closer to the center.

Under these circumstances, unless proposals for changes are introduced in a very carefully controlled and planned manner, the voting process would in all probability lead to rapid movement toward the center.[11] Unfortunately, the convergence need not continue until the absolute center is reached.[12] Close to the center, the area which is preferred to A and is closer to the center, is much smaller than initially. It is therefore more probable than at first that the preferred alternative to A would be farther from the center than A. Cycling becomes more probable. When we get very close to the center, a point randomly selected from among those which could get a majority over the given point would have a good chance of being farther from the center than it is. At this point, however, most voters will feel that new proposals are splitting hairs, and the motion to adjourn will carry.

Discussion of the point is simplified by the use of the "median lines" introduced in the last chapter. As the reader will recall, a median line is a line passing through two individuals' optima and dividing the remaining optima either into two equal "halves" or, if the number of optima is odd, into two groups one of which has one more optimum than the other. Figure XXX shows one such line and a point, A, which is not on the line. If, from point A, we drop a perpendicular to the median line, then the point at the base of the perpendicular, A^1, will be closer to all the points on the other side of the

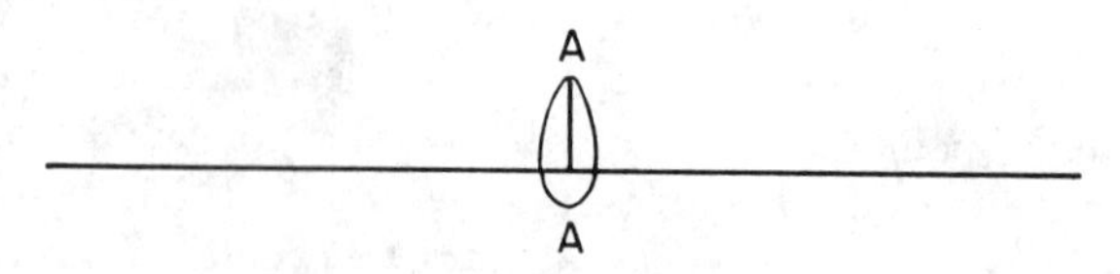

FIGURE XXX

line and the two points on the line than is A. It can, therefore, get a majority over A. Actually there would be a small lozenge, as in Figure XXX, outlining points which could get a majority over A. The geometry of this lozenge, however, will vary somewhat, depending upon the exact location of the individual optima, so we will confine ourselves to the simple perpendicular relationship.

Most of these median lines would intersect in a tiny area in the center of the issue space. If we greatly magnified this area and drew in only a few of the median lines, we would get something which looked like Figure XXXI. If we start with point A, then our theorem indicates that B can get a majority over it. C, on the other hand, can get a majority over B. Similarly, other points can, obviously, get majorities over C. Starting with any point in this general area, it will be possible to select points which will obtain a majority over it. Thus, there is no point which can get a majority over all other points.

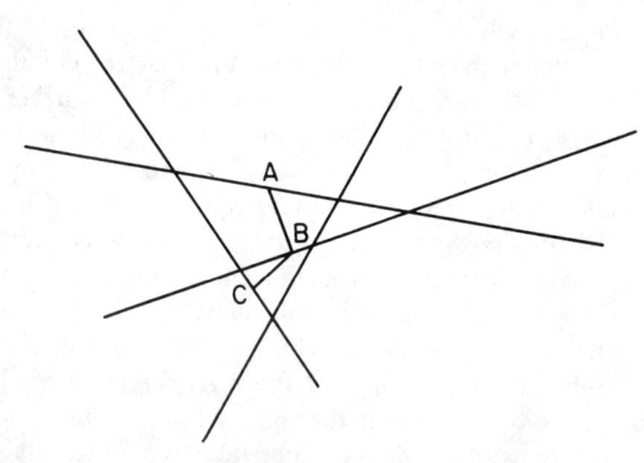

FIGURE XXXI

The area in which the bulk of the bisectors intersect is, of course, very small, but in some cases the point of intersection might be far away from the center of the issue space. Suppose that there are an odd number of points and we select one which is near the extreme outer edge of the issue space. It may be possible to draw through this point two lines, each of which passes through another point and each of which divides the optima so that there is only one more on one side of the line than on the other. The angle between these two lines would be extremely small, but by exaggerating it we get the situation shown in Figure XXXII. Above this pair of

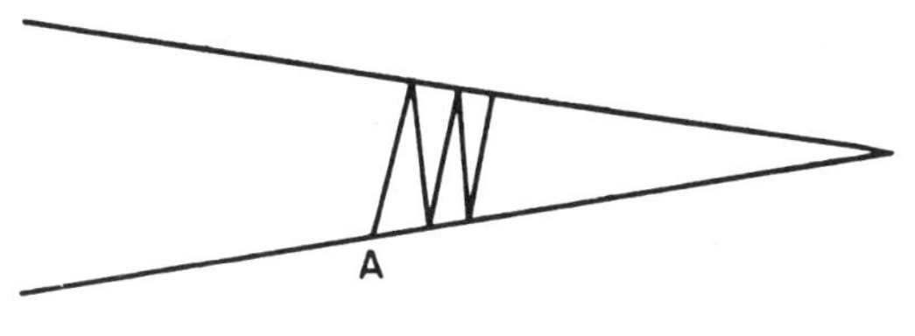

FIGURE XXXII

lines there would be 499,998 optima and below, the same number. The three points lying on the lines make up our total of 999,999. If we start at any point on either of these lines, such as A, we can drop a perpendicular to the other and thus obtain a point which can get a majority over the first point. From this second point, we can then drop a perpendicular to the first line and obtain a point which can get a majority over it. By continuing this process we can eventually approach the intersection point which, by assumption, lies at the outer edge of our space. Thus it is possible, by simple majority voting, to reach points at almost any portion of the issue space. Needless to say, this sort of series of votes is highly unlikely. It can be easily recognized because it would involve a long series of votes in each of which there was only a one-vote difference between the majority and the minority. Since this situation is never seen in the real world, we can feel reasonably confident that this type of movement away from the center does not occur.

Since standard voting procedures do not permit infinitely fine adjustment, the fact that majority voting would not lead to a unique solution seems of very little importance. Black defines a "majority motion" as a proposition "which is able to obtain a simple majority over all of the other motions concerned."[13] The rules of procedure make it unlikely that such a motion will be selected by majority voting. The outcome should be a motion which could not get a simple majority over *all* other motions, but only over those other motions which differ enough so that they can be put against it under the procedural rules. The result is an approximation, but a reasonably satisfactory one. Thus, if there is no true majority motion, if endless cycling were the predicted outcome of efforts to obtain perfect adjustment, this would not change the outcome at all if the cycles would only involve motions proposing such small changes that they could be ruled out of order. Even if the cycles slightly enlarged the area of which the voting system was indeterminate, this would be a trivial defect. Only if the cycles would involve "moves" substantially larger than the minimum permitted by the procedural system would they be a significant problem.

The investigation of the likely size of cycles in the real world can proceed by making assumptions about the distribution of voters and the rules of order and then calculating the likelihood of cycles among motions which differ enough so that they could be voted on, or by observing the real world.[14] There would seem to be two possible explanations for this paucity of examples of the phenomenon. Either it doesn't occur very commonly, which would be in accord with the theoretical considerations given above, or it is hard to detect the presence of cycles even when they are present.

In order to examine the possibility that the shortage of real world examples of cycles is explained, not by their rarity, but by the difficulty of detecting them, let us consider the actual methods of voting used in most representative bodies. Under Robert's Rules, or the innumerable variants which exist, the procedure is quite complicated. We need not examine these rules in detail; a simplified generalization of them will suffice. Let us, therefore, examine the following system. A motion is made to move from the status quo. An amendment to this motion may then be proposed, and various subamendments to the amendment. All of the amendments and subamendments can be regarded as separate proposals. The distinguishing characteristic of this system is that a whole set of proposals is made before any of them are voted upon, and then that they are voted upon in a fixed order which is known in advance.

Suppose that the status quo is A in Figure XXXIII. B is offered as a motion. B′ would be offered only if its sponsor thought it could beat both A and B. (Or, in special circumstances, if it might lead to a blocking cycle. This will be discussed below.) But people do make mistakes. Let us suppose, then, that someone in error offers amendment B′. This is followed by subamendment D, which has been correctly calculated and can beat A, B, and B′. For our pur-

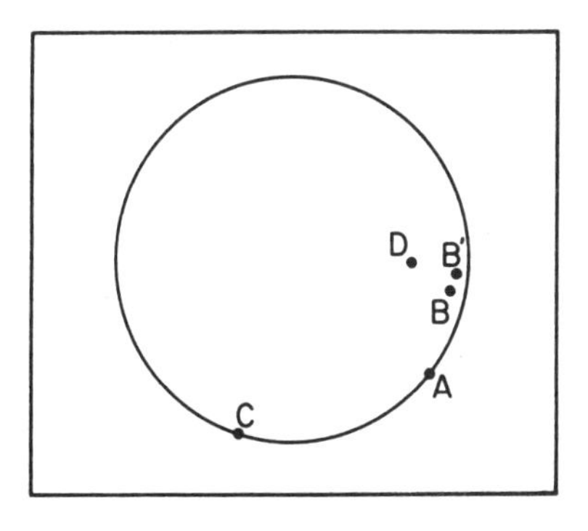

FIGURE XXXIII

poses, we may use a simple set of rules providing that motions, amendments, and subamendments are voted upon in reverse order from that in which they are proposed. Thus, D would be put against B′, would win; would be put against B, and would win; and then would be put against the status quo, A, and would win again. Note that B′, the miscalculation, has no effect on the voting except to delay it slightly. Such mistakes will certainly be made, but they will have no effect on the outcome. We can, therefore, ignore them.

Let us now consider the possibility of cycles. In Figure XXXIII, we start with the status quo at A. Suppose that a motion, B, were introduced which could beat A. Either by accident or by calculation, another motion, C, might be introduced which could beat B, but would be beaten by A. Deliberate contrivance of such a cycle by people who prefer A to B, but realize that B would win in a direct confrontation, would be rational. This existence of a cycle, of course, does not prevent other amendments from being offered. D, for example, could beat any member of the cycle.

If, however, D were not offered, the voting on C, and B, and A would not immediately lead to an apparent cycle because A would be chosen if they were voted on in the order specified. This type of concealed cycle, however, should not lead to a stable result. Once the voting has led to a return to the status quo, further proposals for change would be strongly urged. If there were strict rules forbidding the reintroduction of a measure which had been voted down,[15] then some other, very similar, proposal would be offered. We could expect to see essentially the same series of proposals and amendments retraced again and again. The absence of this kind of repetition in actual legislative practice is evidence that there are few concealed cycles in real world legislative activity.

So far we have been mostly concerned with the situation when there are only two variables and they are continuous. The generalization of the conclusions to multidimensional issue space is obvious, but the effect of shifting to noncontinuous variable may not be. In Figure XXXIV we show a situation where the two vari-

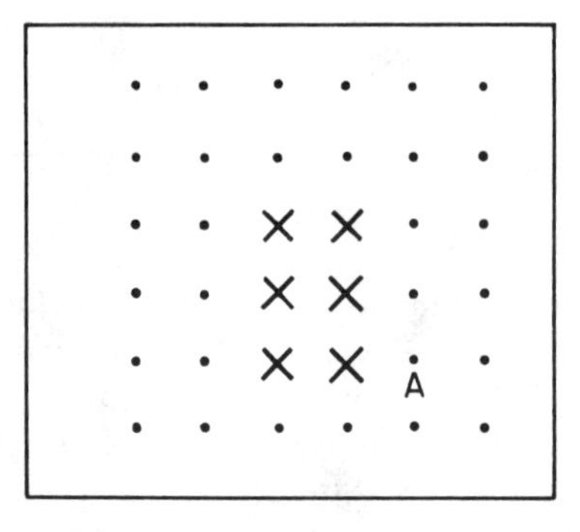

FIGURE XXXIV

ables are both discontinuous, and, hence, only certain points in the issue space are possible. The principal difference which this restriction makes is that, now, far fewer points can get a majority over some given point. Point A, for example, is dominated only by the six points marked with x's. The likelihood that one of the points that dominate A is equally or farther distant from the center than A, is reduced when the total number of points is small, and the likelihood that there will be a set of points which are in a cycle is small. The tiny central area where every point is dominated by some other point is apt to be nonexistent simply because there are too few points in this region. On the other hand, cycles are only unlikely, not impossible. If a cycle does occur, it is likely to be of more than trivial significance if the distances between the points are sizable, and the cycle must, therefore, involve sizable differences of policy. In sum, with discontinuous issue space, cycles are rare but are apt to be important when they occur.

Our discussion so far has been based upon a special type of interdependence of the preference structures of individuals. It is assumed that social states, products, and laws differ in a number of characteristics.[16] Each of these characteristics may be arranged along an axis, either as a continuous variable or as a series of points. Each individual is assumed to have some optimum point in the resulting dimensional space, and it is assumed that the individual's degree of satisfaction falls off as we move away from his optima in any direction. This latter assumption, in the form of perfectly circular indifference curves,[17] is too strong, and we shall shortly demonstrate that a weaker assumption will do as well. Similarly, our assumption of even distribution of the optima over the issue space is a simplifying assumption which will shortly be dropped. Leaving these two issues aside, however, the general picture should raise few objections from economists. Special cases in which these conditions do not hold can be invented, but most choice problems will arise in environments which lead to this sort of preference system. The fact that each person has a preference structure of this sort, together with the fact that they are all in the same hyperspace gives them a rather probable type of interdependence, and our conclusions are essentially derived from this interdependence. Note, however, the rather special form of this interdependence. My preferences do not in any way affect yours. The interdependence comes solely from the fact that we are choosing from among the same set of alternatives, and these alternatives are such that they restrict the form of our preference structures in a way which leads to our conclusion.

So far we have used two unrealistic assumptions, that the indifference curves are all perfect circles and that the individual optima are evenly distributed over the issue space. The elimination of these assumptions will make the model much more realistic. Let us begin by considering more realistic distributions of the optima.[18] Presumably the common distribution is to have the optima arranged in a bell-shaped distribution with its peak somewhere in the issue space. This distribution raises no particular problem for our demonstration. The "median lines" could still be constructed, and they would still mostly intersect near the peak of the distribution. This would mean that the same tendency to move to a small area in the center would exist. Similarly, a skewed normal distribution would raise no particular difficulties, although the area where most median lines intersected would not necessarily be at the peak of the distribution.

Multipeaked distributions raise more difficulties, although the only clear cases leading to significant cycles which I have been able to produce involve extreme degrees of multipeakedness. As an example, if the optima are arranged in three discrete groups arranged in a triangle, as in Figure XXXV, and roughly a third of them are in each group, then cycling would occur over an area of

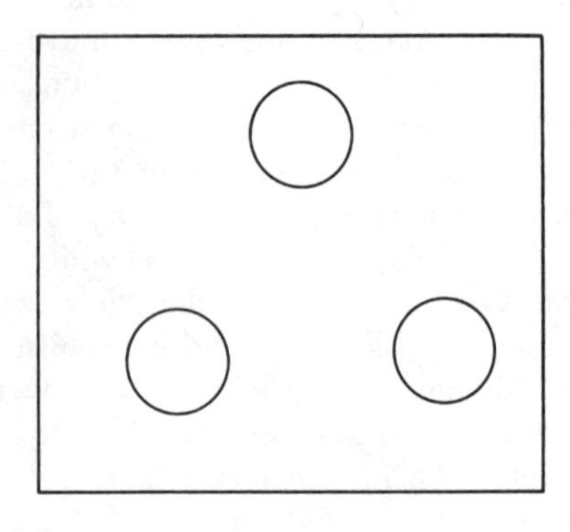

FIGURE XXXV

significant size. The general difficulty of finding such phenomena can, perhaps, be illustrated by the fact that division of the optima into four equal groups does not lead to significant cycles.

The elimination of the assumption that the indifference curves are perfect circles, raises no particular difficulties if large numbers of people are involved in the decision process. Arrow wrote about the problem of selecting a "social welfare function," and this clearly would involve many millions of individuals. In the more general problems of collective choice, there are also normally enough individual preferences to be "aggregated" so that the law of large numbers can be applied. Granted large numbers of individuals,[19] any reasonable preference structure will aggregate in more or less the same way as our perfect circles. On the average, and with large numbers of voters the average is what counts, majority voting will choose the alternative which is closest to the optima of the majority of the voters. Suppose, for example, that we have two groups of 1,000,000 voters, A and B. Assume that of two alternatives X and Y, X lies closer to the optima of all members of group A, and Y closer to the optima of all members of group B. We would expect about 1,000,000 votes for X and about 1,000,000 for Y if that choice were put to the 2,000,000 voters even if 100,000 voters in group A had preference mountains which were so shaped that they preferred Y to X. This would be so because the law of large numbers would indicate, lacking some special phenomenon, that there would also be about 100,000 voters in group B who preferred X to Y.

Note, however, that we only get an approximate result here. With the voters evenly divided, the small random variation which we would expect would decide the election. If the numbers in the two groups differed by more than the likely random variation, however, the outcome would not be affected. Thus, the introduction of what amounts to a stochastic variable, by considering indifference curves which are not perfect circles, blurs our conclusions a little and expands the small area in the center of the distribution in which cycles can occur by a small amount, but doesn't basically alter our conclusions.

That the majority voting process normally leads to a determinate outcome and that this outcome is apt to be reasonably satisfactory will surprise no practical man. Clearly this is what does happen. One of the real problems raised by Arrow's book was why the real world democracies seemed to function fairly well in spite of the logical impossibility of rationally aggregating preferences. The solution I have offered, that no decision process will meet Arrow's criteria perfectly, but that a very common decision process meets them to a very high degree of approximation, permits us to reconcile the theoretical impossibility with the practical success of democracy.

CHAPTER IV

Hotelling and Downs in Two Dimensions

Harold Hotelling, in his early work on monopolistic competition, proposed a model in which two drug stores were seeking optimal locations in a community which lay along a single road. He suggested that this model might, with suitable modifications, be applied to politics. Anthony Downs, in his book *An Economic Theory of Democracy,* made this application and derived from it a number of very realistic conclusions.[1] This model, however, can be subjected to several criticisms. In the first place, although it gives very good predictions for two parties, the Downs's model does not seem to fit very well the situation involving more than two parties or candidates. Second, clearly most political choice situations are not one dimensional, as the Hotelling-Downs model assumes. Finally, the model does not easily deal with the log-rolling situation, or, indeed, any situation in which the voters fall naturally into small, special interest groups. It is the purpose of this chapter to broaden the model to two dimensions. It will be demonstrated that this two-dimensional model gives the same conclusions as Downs's one-dimensional continuum in the special case of two parties or candidates; and that it gives different and more helpful results if there are more than two parties. Only special cases of log-rolling and differential intensity may be shown on two dimensions, but analogical, higher dimensional models may use any number of issue dimensions, and log-rolling may be analyzed with these higher-dimensional models.

Turning to our first objective, Downs's basic assumptions do not exactly fit our present purposes, but they require only very slight alteration. We may assume the following conditions[2]:

1. A single party or candidate is chosen by vote to carry out some set of activities.

2. Elections are held periodically.
3. The number of people eligible to vote is fairly sizable.
4. Each voter casts only one vote. (Analogical models with differential voting can be easily constructed.)
5. The party or candidate who receives the most votes will be given control over the relevant activities.
6. Normally there will be more than one candidate or party in each election.

Given this political system, we can assume voters with some given set of preferences, and examine the outcome. Let us use first the assumption that they are evenly distributed over a two-dimensional issue space, and that their indifference curves are perfect circles about their optima. As in the last chapter, these assumptions are convenient, but not really necessary. As a further condition, I shall assume that all voters always vote for the alternative which stands highest on their preference schedule.[3]

Under these circumstances the geometry developed in the last two chapters can be readily applied. Suppose that a party or candidate, knowing that there will be one and only one opposition party or individual, is contemplating a platform taking some position on Figure XXXVI. If they can get to the exact center, then

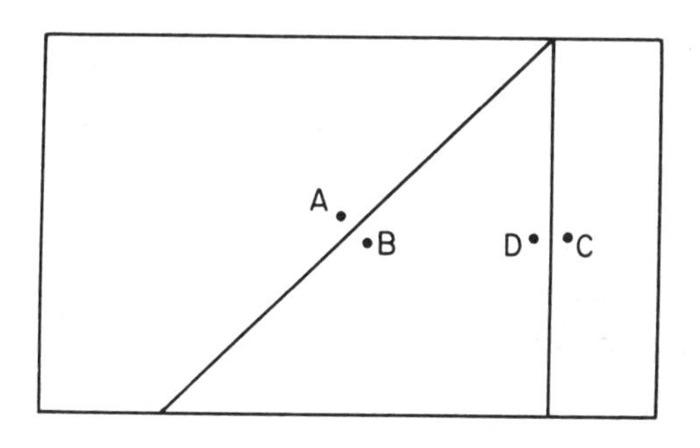

FIGURE XXXVI

they have won, since there is no other point which can get a majority over the center. Errors occur in politics as in other sectors of human life, however, and they may miss. If the first party chooses a point which is not in the exact center, such as A on Figure XXXVI, then the second party has a chance of winning. Probably the information on voter preferences available by the time the second party or candidate makes a choice, will be better than when the first choice was made. The optimal strategy for the second party is easily determined. It should draw a line from point A through the true center of the surface, and then take a position on that line which is on the same side of A as the center and as close to A as is possible without merging into A's position in the view of the voters. B is such a point. The vote for the two parties may be obtained by simply erecting a perpendicular bisector on the line connecting A and B and observing which side has the largest area and, hence, the most votes. In Figure XXXVI, A has chosen well, and there is little difference between the two points in terms of the number of votes they can draw.

Suppose, however, the first choice of position is badly wrong, like C in Figure XXXVI. In this case the same rule applies. The voter-maximizing position, D, is not near the center, but very close to C and on a line connecting C with the center. Thus, an extremist candidate can pull a vote-maximizing opponent far off toward the extremist's desires. These conclusions are exactly those drawn in the *Politics of Bureaucracy*[4] from a two-dimensional model. The reason is simple. If we extend the line connecting the two points chosen, we have a one-dimensional continuum. The votes for each position can be projected upon this continuum, and the conclusions we have drawn will follow from this simplified model.[5]

The model here developed, however, is a vote maximization model. Dr. William Riker in his *The Theory of Political Coalitions*[6] argues that the rational political party would, instead, aim for the minimum winning coalition because the gains per capita are greatest with such a coalition.[7] This situation is shown on Figure XXXVII. If one party makes a mistake and takes position A, then the vote-maximizing reply is position B. All of the voters whose optima lie to the left of line I, will then vote for the second party.

For those voters at the left end of the issue space, however, this is a fairly small gain. B is still a long way from their optimum. If the second party takes position B′, it can still gain a majority of the voters, all those to the left of line II. Further, by almost any definition of "members of the party," much more than a majority of party members would prefer B′. If, for example, we assume that all those who would have voted for the party if it had taken position B are members, then B′ would get almost two-thirds of the votes in

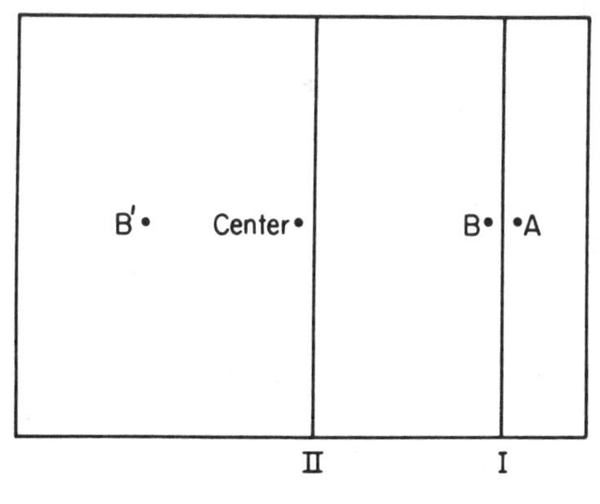

FIGURE XXXVII

an intraparty ballot. The switch from B to B′ is, in essence, a change from a position which gains a little for a large group of people to a position which insures a larger gain for a smaller group.[8] One might hypothesize that a party with a well-established organization of professional politicians in power might choose the B strategy, while a group of young rebels, trying to take over the party, would espouse policy B′. Intraparty revolutions would normally consist of throwing the B clique out. After the B′ clique was firmly established, it would probably drift gradually in the direction of B, and then have trouble with a new group of "Young Turks."

It would probably be possible with little trouble to convert our simple election model into an intraparty contest for control and subsequent election. But it will not be attempted here. One word of warning is perhaps in order. I have placed B′ as far to the left as it can be and still win over A. It could be reasonably assumed that the party or candidate who had originally taken position A would shift leftward if their opponents took a position to the left of B, and, hence, B′ as I have drawn it would probably not win. Before deciding on a position for the second party it would be necessary to make some calculations of the likely response of the first, and this would surely lead to a position closer to the center than B′. Models of this sort would be easily developed, but many different sets of assumptions seem about equally descriptive of the real world, so none will be presented here.

If the two parties have gradually drifted into a position like A and B, then the party with a majority may have an intraparty revolution which will shift the controversy nearer the center of the issue space. Although the details are radically different, something similar might happen within the minority party too. In Figure XXXVIII we assume that the two parties have gotten themselves

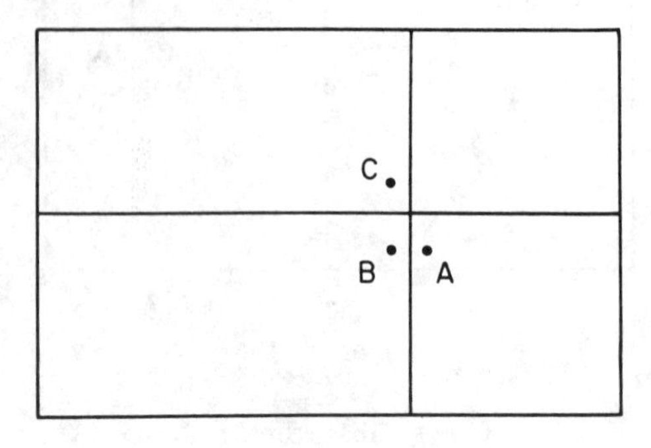

FIGURE XXXVIII

into positions A and B. Obviously, the minority party hasn't much chance. In a strategy carried out by Bryan and Goldwater, a rebel group within the party might propose a drastic reorientation of party strategy. If they can seize control of the party, they might seek to shift to a position such as C which appeals to a different combination of voters than does A. If the strategy is a winning one, only part of the party, defined as those who would have voted for A, will benefit. Again, part of the party is sacrificed, this time in a search for new allies rather than a simple quest for improved returns. Note that the new position of the two parties is not very close to the center, but with this arrangement, movements to the left by either party are likely to be highly profitable. Hence, it can be assumed that they will occur and the eventual location of the parties will be near the center of the issue space. Note also that the space between A and C is greater than that between A and B. This is based on the assumption that the voters will be able to make finer distinctions along the lines of division they have grown used to than upon a new set of issues.

So far, however, we have been using our geometric tool simply to duplicate results already obtained by other methods. When we turn to more than two parties or candidates, it shows a clear superiority over the previous models. In Figure XXXIX a three-party system is considered. Suppose that A and B have adopted positions and a third party C is considering what position

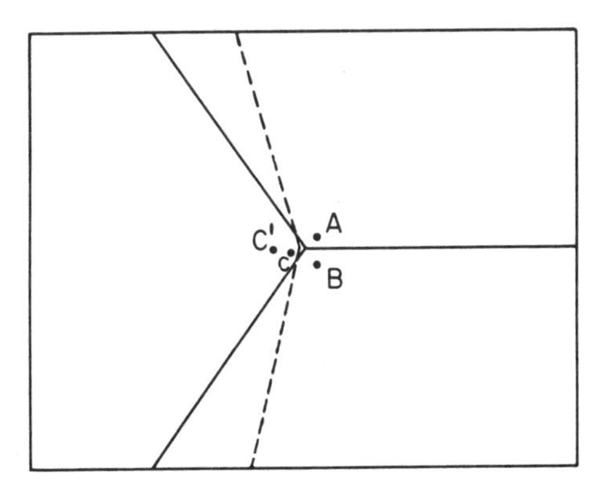

FIGURE XXXIX

it could adopt. Further, suppose that whichever party gets the most votes will be elected. In any effort to deal with this problem in general terms, the corners of the rectangle make problems, and hence it is better to adopt a circular issue space, but for our present purposes we can get by with a rectangular issue space.

If the third party chooses a point which is close to A and B (C, for example), then it will surely lose, as shown by the solid lines which divide the area into the segments closest to each of the three. If, on the other hand, the party chooses a point somewhat further out, C′, then it will win, as the dotted lines show. Of course, it is possible to lose by moving too far out, but, at any rate, the optimum location is not as close as possible to the location of the other two parties. This result, of course, applies also to A and B. If more than two parties or candidates are expected, then the vote-maximizing position is not close to your opponents, but well away from them.[9] In the real world, multiple party systems always involve significant differences among the parties, so our result is realistic.

Previous discussions of multiparty systems have usually been based on a single-dimension, left-right continuum. The assumption that the parties differed along only one axis hardly seems justified. To name but one problem, almost all countries with multiparty systems have some sort of farmer's party which doesn't fit into the left-right structure. Our system, which can be generalized to any number of issue dimensions, has no difficulty in dealing with such interest-oriented groups. Nor would it be difficult to fit in a set of parties organized from left to right, if such a structure were found.

It seems likely, however, that the customary left-right system reflects the analytical tools available, not reality. The rules first adopted to seat the members of the French legislature are surely not universal natural laws. It seems likely that all parties in multiparty systems should be thought of as occupying portions of a multidimensional issue space, which are differentiated by their direction from the center, not by their location upon one particular dimension.

So far I have used my own version of the Hotelling-Downs model to demonstrate that it can be generalized to two or more dimensions. Downs's version is somewhat more complicated than mine, but it also can be generalized in the same way. The principal differences between the models are two. My model assumes that the voter always votes for the alternative which is closest to his optimum; Downs's, that he does so only if the nearest alternative is closer than some crucial distance to his optimum. If it is farther away, he does not vote. With this model the distribution of the voters becomes crucial, and Downs must therefore make realistic assumptions about this distribution, while I can get away with very simplified distributions. There is no particular difficulty, however, in specifying such distributions on our two-dimensional model, and the next few pages will be devoted to duplicating Downs' reasoning, using such models.

On Figure XL a roughly bell-shaped distribution of the voters, comparable to that in Downs' Figure 2 (p. 118), is shown.

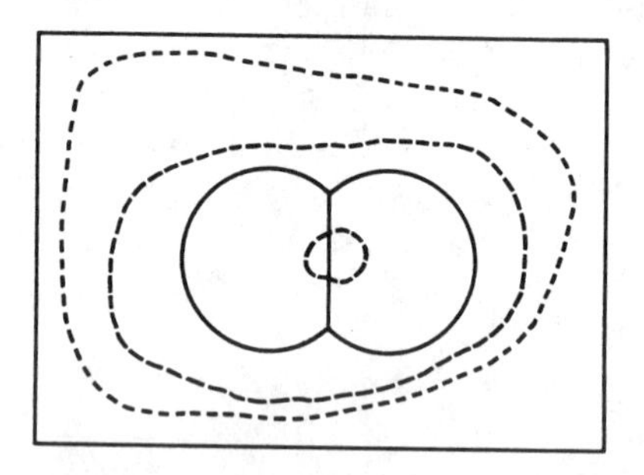

FIGURE XL

The voters are assumed to vote for the nearest alternative as long as that is no more than three-fourths of an inch away. Vote maximization on the part of the two parties leads to positions much like those shown, with the voters within the compass-drawn circles voting for each party. Downs's Figure 3 (p. 119) shows that, with his assumptions, a two-party system can lead to parties which are far apart, if the voters' preferences are arranged properly. Figure XLI duplicates his model in two dimensions. Neither party can

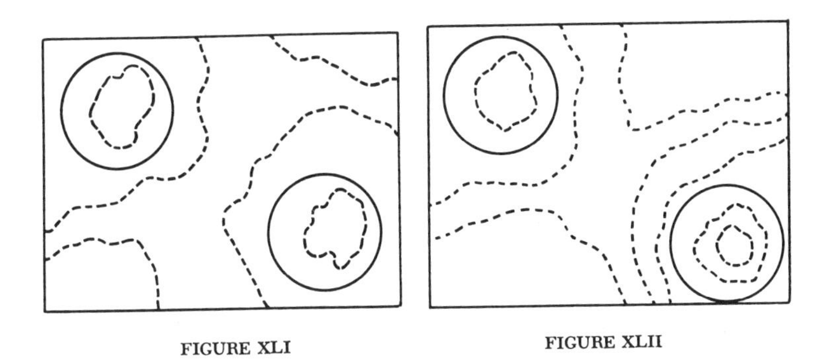

FIGURE XLI FIGURE XLII

gain votes by moving closer to the other, yet they are very far apart. With such a distribution it would be mere coincidence if the hills upon which the two parties sat were of equal size. Downs's Figure 4 (p. 121) shows the situation in which they are not, which is presented in a two-dimensional model in Figure XLII. The party in the upper left-hand corner could hardly expect to win any elections.

Finally, Downs shows on Figure 5 (p. 122) a distribution of voters which he feels will lead to a multiparty system. This distribution could be displayed in two dimensions in a variety of ways, but Figure XLIII shows the simplest distribution. Needless to say, the choice of which assumptions are the more useful is, in the present state of knowledge, necessarily a matter of subjective judgment. It should not, however, be impossible to test which set is more in accord with the real world. The issue distribution of the nonvoters would be quite different in different models, and this could be investigated quite easily. There are, also, numerous more complicated models which could easily be developed. For the time being, however, these models have considerable explanatory and predictive value. Walter Lippmann ran a number of columns explaining that Goldwater would be drawn to the left by more or less similar forces, and the sharp shift to the right by Johnson after Goldwater's nomination is a further demonstration of their power.

Turning now to the problem of log-rolling, only peculiar and

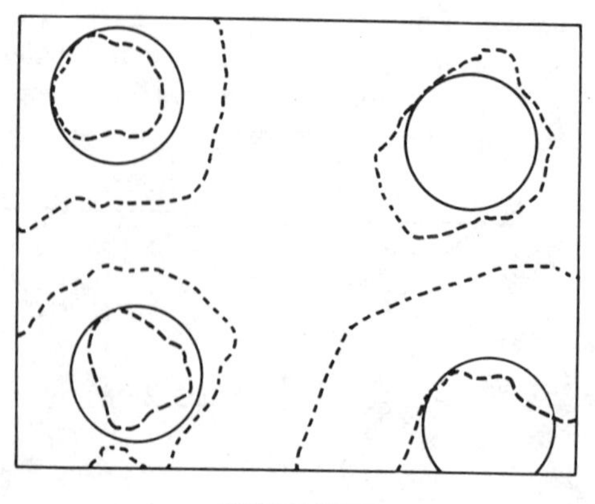

FIGURE XLIII

special cases can be shown on a two-dimensional issue space. Even for these cases we will have to represent the preferences of interest groups, not individuals. Interest groups, of course, are built up out of individuals, and normally do not represent a group of people with identical preferences, but people who feel strongly on one issue and less strongly on others. Their individual optima, both on the dimension in which they feel strongly and on the others, may vary, but the ultimate effect is an approximation of the interest group preferences, which we will use in our analysis. Figure XLIV shows such a set of group preferences arranged to "explain" the lengthy delay in providing federal funds for local schools. It is assumed that the population fell in three rough groups. The optima of the Catholics who favored school aid but who were against such

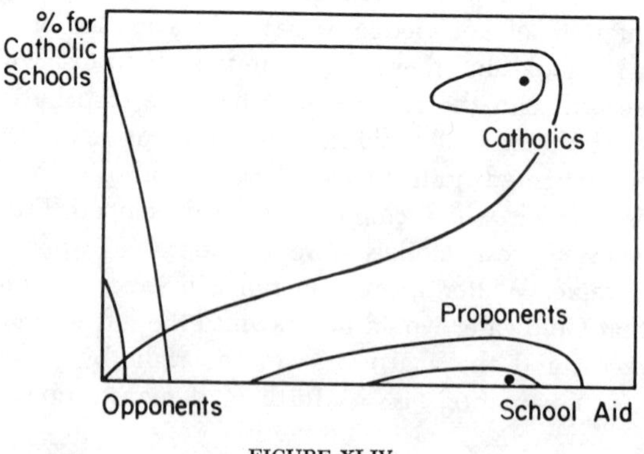

FIGURE XLIV

aid if they were excluded is shown near the upper right hand corner. The opponents of the aid program generally wished to include the Catholics if there was to be such a program. Their optimum was at the origin. The proponents of the program, on the other hand, were normally strongly opposed to any aid to Catholic schools. Their optimum is far out on the horizontal axis. The result, the defeat of the proposal, can be thought of as the result of a deal between the Catholics and the opponents, and thus as representing log-rolling, but this is straining the concept somewhat. Similarly, if we have three persons on a two-dimensional issue space, an agreement among two of them to vote in some point along the contract locus connecting their optima, might be thought of as log-rolling.

Log-rolling, essentially, involves one group of voters or representatives making a trade with one or more other groups under which, say, group A, which favors measure X and opposes measure Y, agrees with group B, which favors Y and opposes X, to vote for Y in return for group B's vote for X. They agree to vote against their own preferences in return for compensation, and this, obviously, is only possible if their desires are of different intensity. Simple log-rolling involves such trades over a sequence of votes. Thus, X would be voted upon and then Y. Implicit log-rolling occurs when the various issues are made up into a large, combined project, usually called a political platform. Thus, a party will run for office on a program calling for both X and Y, and both groups A and B will support it.

The reasons for our inability to present this situation directly on a two-dimensional graph will become clear from an examination of Figure XLV. A and B are shown with their optima and indiffer-

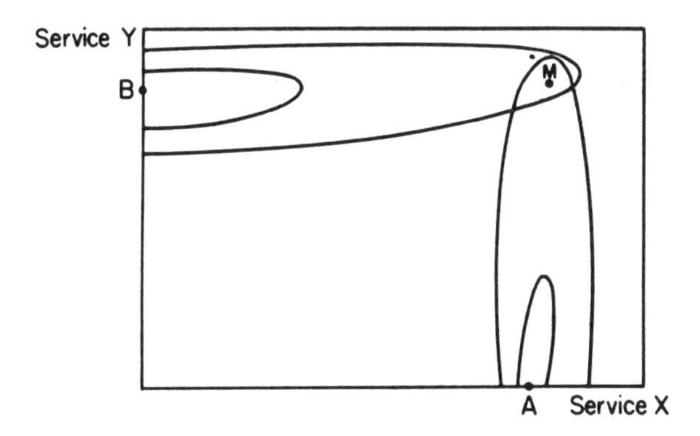

FIGURE XLV

ence curves. The shape of the indifference curves, of course, indicates their much greater interest in X and Y, respectively, than in Y and X.[10] Assuming that these two groups are the total electorate, then there is no need for a bargain unless A and B happen to have exactly the same number of voters. If either were to have even one more vote than the other, then the larger group could simply impose its desires by majority voting and ignore the wishes of the other. Only if the two groups were of identical size, would a compromise on point M be sensible. If, however, there are other voters, say a set C who favors measure Z but is opposed to X and Y (and if the A and B voters also oppose Z), would we expect such a result? If neither A nor B has a majority, but if the two of them make up such a majority, then agreement on point M becomes sensible. A party offering M, combined with strenuous opposition to Z, would get the votes of groups A and B and win.

This situation, however, cannot be fully represented in two dimensions. Z should be placed along a third axis sticking vertically out from the paper. At best, Figure XLV can be thought of as a cross section of the three-dimensional figure at the point where Z is zero. If we think of the matter in this way, then the indifference curves of group C would roughly be segments of circles centering on the origin. Our diagram, then, is not determinant. It helps us to understand, but it does not show the entire situation because the possibility of agreements between groups A and C, or B and C, does not appear on this cross section. Obviously, the real-life situation with many issue dimensions cannot be put on the two-dimensional continuum.

How, then, do we explain the apparent close fit between such two-dimensional models and the complex multidimensional world in which log-rolling is the norm? Clearly the answer lies in the fact that the two types of space are identical to the political candidate or party. If we choose a position in our two-dimensional space, movement in any direction will result in some voters being pleased and others disappointed. If there is another candidate or party, the problem is to choose a position which is preferred by a majority of the voters. If there is more than one party, a position which will maximize votes must still be chosen. We can think of the party as moving through the issue space in search of its optimum locations. In its motion it will continually pick up voters and lose others. The exact shape of the indifference curves of the individual voters is of little interest to it. Since the principal difference between log-rolling and voting for other motives lies in the shape of the indifference curves, with the individual feeling very intensely about some subjects in log-rolling, the log-rolling and nonlog-rolling models are more or less indistinguishable from the standpoint of our hypothetical model party. The conclusions drawn from two-dimensional space without log-rolling are, thus, fully transferrable to a multidimensional Hilbert space which depends largely upon log-rolling. Although the two models are greatly different, the difference is not relevant to the party seeking an optimal position. It follows the same rules of action and ends up in equivalent positions in both situations.

CHAPTER V

Single Peaks and Monopolistic Competition

So far this book has been concerned with the application of economic tools of analysis to political problems. In this chapter, a partial repayment will be made to economics. Duncan Black developed the single-peak preference curve as an application of economic methods to politics; I hope to demonstrate that this tool is of great use also to economists. Specifically, a simple and elegant way of dealing with monopolistic competition is provided. This model will not add very much to what is already known about this field of economics, but it will make presentations to students easier. It will also permit the deduction of several statements about the real world which are empirically testable. Thus, the principal objection of the Chicago School to the whole concept of monopolistic competition will be overcome. Monopolistic competition has always seemed much more realistic, as a description of reality, than either classical pure competition or monopoly. It offers, however, analytical difficulties which have so far sharply limited its use. It is hoped that the single-peak approach will make at least a start in overcoming these analytical problems, and thus make this "realistic" approach of greater technical utility. This model will not, however, be completely without utility to the political scientist. In the next chapter it will be used to analyze opinion formation, clearly a political problem.

At first we will use a simple model of drugstores located in a community which lies entirely along a single road. This will be generalized to more realistic population and location assumptions later.[1] Finally, we shall consider monopolistic competition in which the products are differentiated by physical characteristics, not by the location at which they are sold. In general, the result will, even with the strict one-dimensional model, be highly realistic. Further, we shall see that monopolistic competition is related to a political process by more than the possibility of analyzing both with the same model. Monopolistic competition can be seen as a collective decision process in which the group makes decisions as to what should be produced and the methods to be used in producing it. Further, monopolistic competition resembles voting procedures in that the "decision rule" permits injury to be inflicted on persons who have no power to stop an action. A decision on the location of a drugstore will affect persons other than the druggist and his potential customers. As in political systems, however, the long run effect for every individual is better, even though he will be hurt occasionally, then could be obtained by giving each individual a veto over all change.

But these conclusions can only be obtained by a rather lengthy chain of reasoning. Suppose then that we have a community which is entirely built upon one road, and we are contemplating opening a drugstore.[2] Our problem is choosing an optimum location. Let us begin by assuming that there are no other drugstores in the community, and that anyone who does not deal with us must either purchase his toothpaste by mail, at some inconvenience, or go without. Under these circumstances, a given potential customer, say A on Figure XLVI, will have the demand schedule shown. In

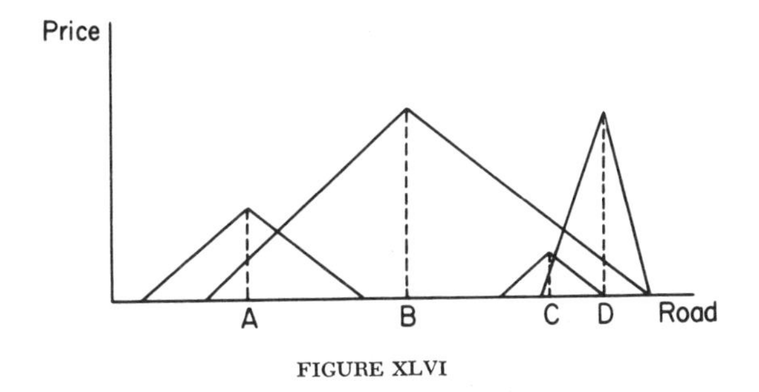

FIGURE XLVI

Figure XLVI the vertical axis represents the price charged. The customer, A, lives at the point indicated on the road, and the distances he would go in order to purchase toothpaste at various prices are shown by the inverted B. Similarly, B, who has a greater desire for toothpaste, and C, who has a lesser desire, would go the indicated distances to buy toothpaste at various prices. D has rather a high demand for toothpaste, but is extremely lazy, hence his demand V is high but narrow. Of course, there are prices at which these customers would not buy, regardless of the location of the drugstore, and there are locations where it would be necessary to pay them to come and get the toothpaste.

On the whole, the variations in the distance that the various customers could be expected to be willing to go to buy toothpaste at various prices would average out from the standpoint of the potential druggist. He is planning to sell to many customers and for this purpose can largely base his calculations on a sort of average demand V for them. Using this average figure, he can draw a similar diagram, illustrated on Figure XLVII, which shows the farthest distances at which a store at a given location can attract customers at various prices for toothpaste. If we assume that poten-

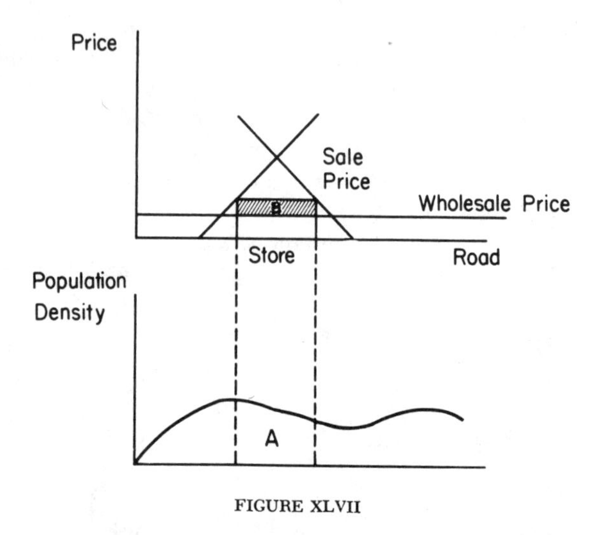

FIGURE XLVII

tial customers are evenly distributed along the road, then only the upper half of the figure is needed to solve the problem. The drugstore can buy the toothpaste at some wholesale price, as shown. Its sale price will determine how far its customers come to buy, and hence how many customers it will have. The rectangle B represents the gross profit on its total sale of toothpaste. The man planning a drugstore should choose his location and price so as to maximize its area. Given that there are no competing drugstores, and that the population is evenly distributed, any location which is not very close to the ends of the road is much like any other, and the only problem is choosing the maximizing price.

If the population is not evenly distributed, however, the

problem is more difficult. In the bottom half of Figure XLVII, a population distribution is presented. If the store is located at the point indicated and the price shown is chosen, then the gross profit will be represented by a three-dimensional space, the base of which is the rectangle B and the front and rear side of which will be the irregular area A. It is immediately obvious that the store owner has chosen a less than optimal location for his store. Another location to its left would bring in more money. That our reasoning to this point can easily be generalized to any spatial arrangement of customers is obvious.

But, suppose that our potential druggist is not the first to arrive in a new area; suppose there are already other druggists established in the community. Suppose, further, that these earlier arrivals have carefully spaced their drugstores far enough apart so that they do not directly affect each other, but that there is no longer room for a new drugstore which will not be in direct competition with another. In Figure XLVIII the situation (assuming customers are evenly distributed) is diagrammed. The customers, or at least some of them, now have more than one pair of alternatives. In addition to the choice of buying or not buying that they had before, they now can choose where to buy. The already existing drugstores have locations and prices, shown as dots on the diagram, and we can start our analysis by assuming that these are fixed. The inverted V which defines the distance that customers would come at different prices, must now be drawn so as to leave out some customers who would come if the competing drugstores did not exist. Further, its sides will not have the same slant. Although the potential drugstore proprietor would still be maximizing the profit rectangle, it is easier to understand his problem if we use a slightly different diagram. In Figure XLIX the locations and price of two existing stores are shown. The potential

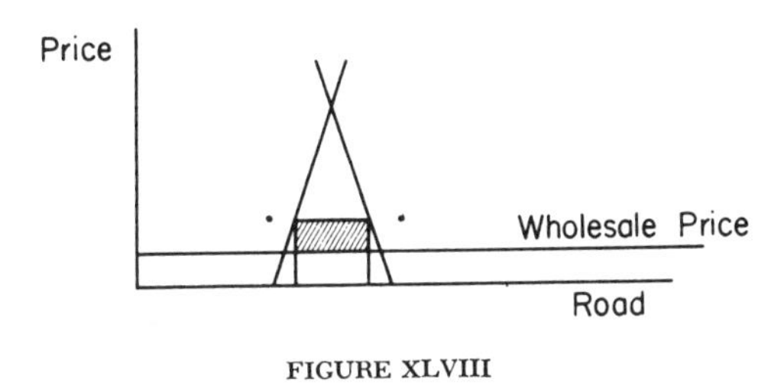

FIGURE XLVIII

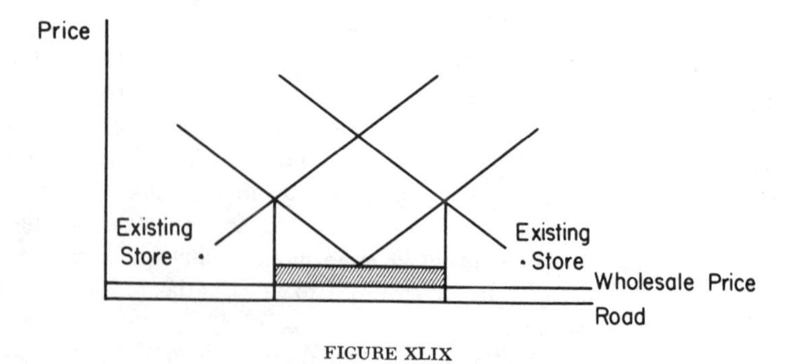

FIGURE XLIX

drugstore proprietor chooses a location and price that will, he hopes, give him the maximum profit. Suppose he wishes to determine the profit he will make. The procedure is simple. Lines are drawn through the location and price of the existing stores with the same slant as those in Figure XLVII. This slant represents the implicit exchange ratio between the distance traveled and the price for the average customer. Each customer's "real price" is the sum of the money price and the inconvenience of travel, and is represented by the height of this line at his location. Lines are now drawn through the proposed price and location, as shown, and from the intersections of the lines, verticals are dropped to the road.

These verticals divide the customers into those who will trade at the new store and those who will continue to trade with the original suppliers. These two verticals are connected by a horizontal line at the proposed price, and the resulting shaded rectangle is the gross profit. The potential druggist would simply try to choose the combination of price and location which maximized this area.

Predictably, the establishment of this drugstore in this area, creating competition with two others, would most assuredly lead the others to cut their prices. We can redraw their inverted V's, showing the new drugstore and its price affecting their geometry, and the gross profit-maximizing price will be considerably lower than previously. This change in prices is, of course, predictable in direction.[3] The man contemplating a new store would surely take it into account. He would make a guess as to what the price charged by these drugstores would be and draw his diagram through these new, reduced prices.

Note that, if the newcomer decides to locate at the point

shown on Figure XLIX, he will reduce the prices charged not only to his own customers but for customers of the two adjacent drugstores who are much too far from his store to ever contemplate using it. Thus, he inflicts a real injury on the two drugstore owners and confers a real benefit on their customers.[4] His decision on location is emphatically not something which concerns only himself and his new customers. We can say, however, that the benefit conferred on the customers and the new owner exceeds the injury inflicted on the two older drugstores. What has happened is that two monopolies have been forced to reduce their prices to a new level closer to a theoretical competitive price, and this will always benefit the customers more than it injures the supplier.

In order to decide upon the optimum location for his new store, the potential druggist should construct these diagrams and carry out the calculations for all potential locations. He, then, simply chooses the one with the highest gross profit. Once he has established his new drugstore, a further potential druggist will have to begin the process all over again, with the new drugstore included among those which affect the size and shape of his V's. In general, the only response of the established drugstores to the opening of a new competitor is to reduce their prices. They cannot be continually shifting to new locations, although in the long run such shifts would be expected.

In our present model, new drugstores would continue to be established until there would be one for every potential customer. This is a consequence of our extremely crude cost function. In the real world, the costs of a retail store are not completely independent of the size of the establishment. To deal with this factor, let us again assume that the potential customers are equally distributed along the street. In Figure L we have shown an inverted V representing the customers who could be expected to come to a given drugstore at different prices for toothpaste, and the average costs which the drugstore would incur with these varying numbers of customers. The potential druggist should try to choose a location and price to maximize the rectangle shown. By bisecting this figure along the location of the store, we can get a standard diagram showing the maximization of monopoly profit.

With this apparatus the potential druggist now can decide that no combination of location and price will give him the normal rate of profit, and, hence, that he should not open a drugstore anywhere. This being so, a stage has been reached where it is no longer profitable to open a new drugstore and the system in our

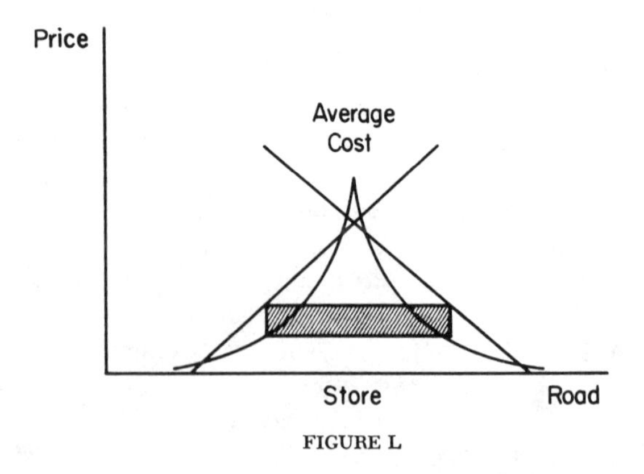

FIGURE L

function, is in a state of equilibrium. In practice, of course, although drugstores cannot easily and quickly shift their locations to respond to the opening of new stores, they do make such shifts over longer periods of time. Thus, even after new drugstores were no longer being opened, stores would occasionally move. If the customers were equally distributed along the road, the end product of this shifting process would be for the drugstores to be distributed at equal intervals along it.[5] This would both maximize the profit of the stores and minimize the *average* man's net cost of toothpaste purchases, taking both the price and the distance traveled into account (again ignoring those at the extremes).

Note, however, that although we can make these statements about the distribution of the drugstores, we cannot say how many there will be. We can readily imagine a situation in which there are seven drugstores placed at equal intervals along our road and in which the interval between each pair of them was not wide enough for a new drugstore to come in, yet where eight, equally distributed drug stores could all be profitable. The historical development of the system is highly important here. The more solvent drugstores, the better, however. Each additional drugstore reduces both the price and the distance the average customer must go to get his toothpaste. Granted that there are economies of scale, and that the customers are not all located at one point, the prices will never get down to the level of perfect competition. The customers can hardly complain about this since it arises out of an effort to please them. Their unwillingness to go long distances to take advantage of small differences in price gives the monopoly power to the drugstore. For them the price is really the sum of the money they pay the drugstore and the "cost" of getting to the drugstore. A competitive price, which would entail fewer drugstores, would really be felt by the average customer as a rise in the total cost of obtaining toothpaste.[6]

Traditional monopolistic competition theory has argued that in the situation we are now discussing there would be an overinvestment in drugstores. It has been pointed out that the drugstores are not operating on the low point of their average cost curves, and that the prices are higher than would prevail if perfect competition reigned. This, however, involves ignoring the preferences of the consumer, and hence is a most unusual type of optimum. If the consumer were indifferent to the distance he traveled to buy his toothpaste, then perfect competition would prevail even if the drugstores were not all at the same locations. Since the consumer is not indifferent and since he does have some sort of trade-off between price and distance traveled, clearly the minimization of the total inconvenience of the travel and the payment should be aimed at.

This requires some new sort of optimum. The firms will be operating at levels which imply at least some monopolization, but the customers will be better off than if the firms were forced to operate at competitive levels. If we define the economic optimum as the point where a retail store's average long-run costs are at their minimum, then it is likely that all retail establishments will always operate at a nonoptimal level. Casual observation would suggest that the limits on the size of stores are not the result of rising cost curves for larger establishments, but the result of the limitations on the distance that customers are willing to come. This would explain the notorious fact that retail establishments are much more concerned with location than any other form of enterprise which is not based on a geographically limited natural resource.

In order to decide what the optimum would be under these conditions, let us consider our community and vary the number of drugstores. The number in existence at any given time is, in part, a function of the particular history of the community, but we can assume different histories and, hence, different numbers of stores. Let us assume, then, that the equilibrium number of stores, when they are all equally spaced and there is no incentive for anyone to start a new one, may vary between thirty and forty.

As the number of stores increases, the average distance that the customer must travel goes down, as well as the price he must pay. This means, clearly, that the profitability of the stores must decline. At forty stores we reach a point where no more stores are possible; forty-one, equally spaced drugstores would all operate at a loss. At thirty, on the other hand, the stores would all be highly profitable. Obviously, if one of the thirty were sold, the price would include a fee for its locational monopoly, while if one of the forty were sold, the price would just cover its real assets. In the first case the capital value of the store would be greater than the total of real investments in it, while in the second it would not. For our purposes the second situation is the optimum. Needless to say, everyone will be as well or better off in this situation than they would be if the traditional optima, with its simple disregard of the customer's convenience, were accepted.

The point is perhaps easier to understand if we consider the small ice-cream trucks. These trucks, unlike the drugstores, are instantly mobile, and can respond to competition not only by price changes but also by changing their locations. Suppose, then, that thirty of them were equally spaced along the single road of our community. If we keep the same assumptions about the distance that customers will travel as we have used for the drugstores, it is clear that no new ice-cream truck would be brought in if its owner thought the other trucks would hold their positions. However, obviously, if the new ice cream truck moved between two existing trucks, all three trucks would operate at a loss. The two original trucks would respond to this by moving a short distance away, and this would set in motion a chain reaction which would end with thirty-one ice cream trucks equally spaced along the road.

Each of these thirty-one ice-cream trucks would now be earning more than the market rate of return on the resources committed to it. The opportunity would therefore exist for a thirty-second ice-cream truck to be put into the system. Once again this would set in motion a series of readjustments. The process could be continued until there were forty ice-cream trucks evenly distributed along the road. Each time a truck was added, the average distance that individuals had to walk and the price they had to pay would go down, and the profits of all existing ice-cream trucks also would go down. The final equilibrium with forty trucks would only occur when the addition of another truck would result in losses for all forty-one trucks. Note, however, that the prices being charged by the forty trucks would still include a monopoly element. There would be no monopoly profit, because no extraordinary profits would be obtained, but each ice-cream truck would be operating inside the point where marginal cost and price are equal. This is the situation which is normally characterized as representing an overinvestment in the industry.

In fact, the situation is Pareto optimal. Consider the change which takes place as each ice-cream truck is added to the system, assuming, of course, that the addition is profitable to its owner. The movement of all the trucks will mean that some people will have to walk a greater distance to get their ice cream than before, but the average distance is reduced, and clearly those who gain could easily compensate those who lose. The price after the change will still be above the "competitive level," but it will be a little lower than before. It is elementary that if a monopolist is in some way forced to reduce his price, the benefit to his customers is greater than the injury to him, and, consequently, that the customers could compensate the producers.[7] Thus the set of changes set in motion by an additional truck is Pareto optimal, until the addition of another truck would cause losses. Hence it would appear that each additional truck constitutes a Pareto optimal change, and the point where an additional truck would lose money is the Paretian frontier.

According to most discussions of monopolistic competition, however, this is not true. The argument is that the true Pareto optimum involves fewer trucks, operating at full capacity and charging lower prices with no element of monopoly present. This situation, of course, is not obtainable by market means. If it were possible for a truck to get normal returns by charging the competitive price and attracting customers from far enough away to get the full use of his equipment, then this is what would happen in the normal course of events. For monopolistic competition to exist, the difference between the monopolistically competitive price and the purely competitive price must be less than would be necessary to attract the additional customers necessary to support an enterprise at competitive prices.

This, however, does not prove that our monopolistic competition optimum is Pareto optimal. At the optimum which we have defined, all enterprises are on the declining portion of their average cost curves. Under these circumstances, it is always possible to construct a system of differentiated prices which will give to everyone a lower price than would be charged with a single price, while bringing the same profit to the owner. If we imagine the customers of one of the ice-cream trucks founding a cooperative, buying the truck, and setting up a system under which people who had previously lived too far away to buy ice cream at that truck are attracted by a set of differentially lower prices, with the farthest customer getting the lowest price, it is clear that they could improve on the market organization. They would expand their system of differentiated prices until the reduction in the average cost obtained from the addition of one more customer was equal to the additional cost imposed on that customer by coming to the cooperative ice-cream truck instead of its nearest competitor.

Thus, we find three price structures: the monopolistic competition optimum, the "Coop" optimum, and the "competitive level." In the real world only one of these, the monopolistic competition optimum is obtainable. It would clearly be impossible for anyone to compute the system of differentiated prices necessary to reach the "Coop" optimum, and the competitive level is not only incomputable but also undesirable. This is not, however, a real disadvantage. If we widen the scope of the factors which are taken into account, the monopolistic competition optimum will assuredly be Pareto optimum. Traditional economic theory ignores the costs involved in reaching agreements. For most economic problems this is sensible, since such costs are very low. When, however, we consider collective agreements involving many persons, the costs of obtaining agreement may be too large to be ignored. In this case a considerable number of people would have to enter into an arrangement in which different people would be charged different amounts. This is the kind of situation in which it is sensible to invest resources in strategic bargaining to obtain an especially favorable price for yourself. When strategic bargaining on the part of many people becomes sensible, it is normally true that the costs involved in this bargaining are so great that they will outweigh even quite substantial benefits which would otherwise arise. The normal way of dealing with this problem is to agree upon a decision rule, say majority voting, which does not require unanimous consent and hence eliminates the possibility of excessive strategic bargaining.

Such a decision rule, however, will also not reach Pareto optimality in each individual decision. The rule itself may be Pareto optimal in the sense that it is the best for everyone. Taking everything, including the costs of organization and bargaining, into account, it may give each individual the maximum present discounted value over all future decisions although not one of those future decisions will be Pareto optimal. We may search for a Pareto optimal rule for social organization in monopolistic competition in this larger sense; almost certainly this rule would be to leave the situation alone. The costs imposed on the individuals by the lack of

Paretian optimality in each case would be more than outbalanced by the costs involved in organizing any other way of dealing with the problem. Until we get methods of measuring these costs, we cannot say this with certainty, but it certainly seems reasonable.

As we have seen, however, monopolistic competition in the short run will reach an equilibrium which is not necessarily the optimum. If the stores are immobile, and most stores are, then they may be spaced far enough apart so that there is a monopolistic profit to be gained, but close enough together so that new stores located between any pair would be unprofitable. This situation, also, may be Pareto optimal in the larger sense which includes decision costs. At any event, there is no reason to believe that this situation involves overinvestment in the industry. Clearly the investment, in real factors here, may be less than optimal. The capitalized value of the monopoly may be important, but this does not really involve saving. It comes, without any expenditure of resources, as part of the property. Theoretically, forcing the stores to lower their prices until there was no monopoly profit left would be desirable, but the costs of organizing such a move would probably be extremely great. For one thing, we would have no way of telling when the individual managers stopped trying to be efficient, since the price would guarantee them a reasonable return on their assets anyway.

So far we have been dealing with a mythical community in which the population is evenly distributed along a road. We can get closer to the real world by dropping the even-distribution assumption. If we consider our model store in which the average costs decline as the number of customers increase, it is immediately obvious that the denser the population, the larger the individual store, the closer together they are, and the lower the price they charge. These, of course, are testable propositions.

However, an important modification must be made. We have assumed that costs of transportation are essentially constant. In the real world this is not so. If we move from very sparsely settled areas toward the center of a large city, at first the cost of moving a given distance goes down a little, as the road net improves. Then it remains constant until the population gets dense enough to raise congestion problems. As we move into the center of a city, the costs of getting to a drugstore sharply increase. Instead of driving to a large parking lot in front of the drugstore, we will normally have to walk, or at least park some distance away and walk. This means that transportation costs once again rise, and we might expect a large number of small drugstores conveniently located, instead of a few giants. The details would be complicated, but here we have a fertile field for empirical investigation. Measures of costs and correlation between these costs and the size, density, and prices of drugstores would be both possible and extremely interesting.

We can get closer to the real world by considering varying population densities. For this purpose let us return to our ice-cream truck model, which is the classical Chamberlin type, and assume that we are interested solely in the ultimate equilibrium, not in the situation when there are just a few ice-cream trucks far enough apart not to bother each other. In this situation the owner of a truck must take both the location and price of his nearest competitor on each side into account. Suppose an ice-cream truck is located at A on Figure LI and the trucks on each side are located at B and C, charging the prices shown. The lines drawn through B's and C's prices represent the total cost (price plus transportation) for customers at different locations purchasing their ice cream from these trucks.

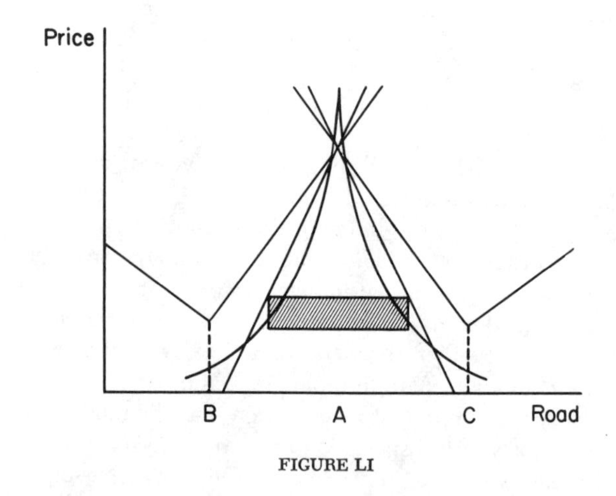

FIGURE LI

If A charged the same price as B, the customers who were closer to A's location would trade with him, and those closer to B's would trade with B. If A charged a higher price than B, the customers would trade with B as long as the sum of his price plus transportation was less than the similar sum for A. The inner triangle in Figure LI shows the distance from which A could expect to attract customers at each price. As before, he would seek the profit-maximizing price, considering his costs. This would involve maximizing the shaded rectangle. If, however, there was any true entrepreneurial profit, the other ice-cream trucks would be attracted, and eventually they would be dense enough that the rectangle would shrink to a line. This situation, the Chamberlin equilibrium, would occur when the average cost lines and the two sides of the triangle were tangent.

In order to make the analysis easy, let us cut Figure LI in half along a vertical through A so that we need only consider distances in one direction from the ice-cream truck. This gives us Figure LII, in which the curves are drawn as tangents. For a cost curve, I have used the simple fixed cost plus constant marginal cost model, with the equation $DWP = 10 + DW$. W represents distance along the horizontal axis representing a road. D represents the customer's density along the road; hence, DW represents the number of customers. In our very simple system where each customer buys one unit, this is equivalent to sales. The vertical axis, of course, represents price.

FIGURE LII

The demand line, $2W + P = A$, simply shows the distance from which customers would come at various prices. Since this is not altered by the density of population, there is no D in this equation. Solving these two equations for P, we get

$P = 10/WD + 1$ and $P = A - 2W$. Differentiating and setting the differentials equal to each other to get the points of tangency results in $-10/DW^2 = -2$. Solving for W, we got $W = \sqrt{5/D}$, which by substitution gives $P = 10/D \sqrt{5}/D + 10$. For any value of D, we can now compute the price charged and the equilibrium distance between ice-cream trucks. This is shown geometrically on Figure LIII. The size of the business (measured in customers),

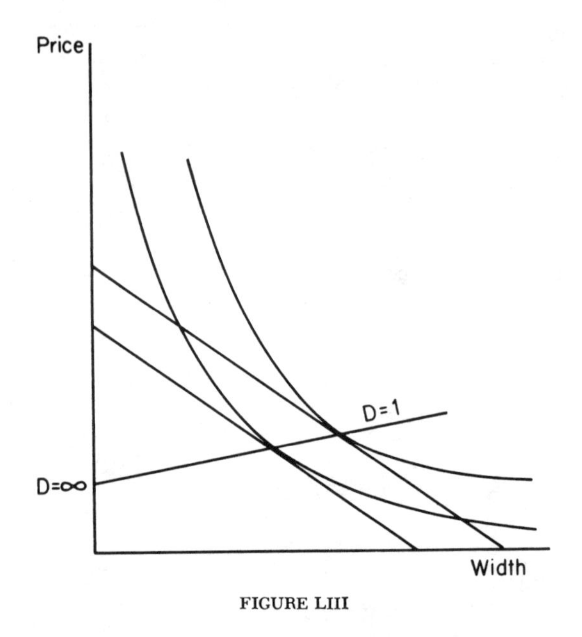

FIGURE LIII

given by DW, is also readily computable. Note that as density of population increases, the price and distance between trucks decrease, while the size of individual installations goes up.

This is, of course, the rapid adjustment equilibrium, not the slow adjustment equilibrium which is the common case in real life. Nevertheless, these conclusions would approximate the slow adjustment equilibrium as well. This means that we have a set of empirically verifiable hypotheses about the real world. One of the principal objections offered to the monopolistic competition system has been that it is not testable.[8] It would seem that this objection, at least, must be given up. Unfortunately, as we will see, a further set of hypotheses about the real world can also be deduced from this model, and these hypotheses have a tendency to offset our first three. Thus, a test would involve considerable care in keeping the two sets of hypotheses separate.

Suppose that something happens which changes the transportation function. Throughout most of modern history, transportation has been getting easier, and this should be reflected in the structure of our economy. On Figure LIV (which should be fairly familiar to all economists) two "demand" lines are shown, each representing a particular "cost" of transportation. The less steep one, of course, represents the cheaper transportation because the cost of going a given distance is reduced. Assuming that the cost structure is not changed, the point of tangency moves down to the right as the cost of transportation goes down. Thus, the equilibrium price falls, size increases, and distance between ice-cream trucks increases as transportation gets easier for the customers. This is, of course, in accord with historic experience, but so many other things were happening at the same time,[9] that this can hardly be taken as a test of the hypothesis.

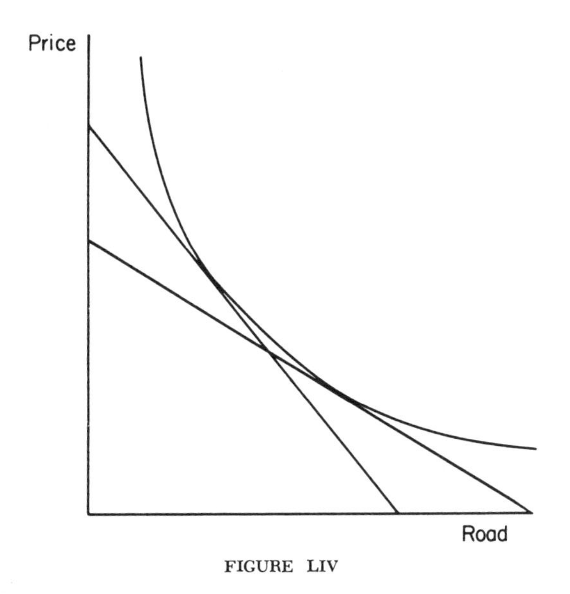

FIGURE LIV

If costs of transportation increase, just the opposite effects should be expected. This can be fairly easily observed in the present-day world. If we start in a desert in Nevada and work our way to higher and higher population densities, ending up in New

York City, transportation costs will remain more or less constant, with perhaps some slight decline, until the population density gets high enough so that congestion begins to make travel inconvenient. In downtown Manhattan, transportation costs are very much greater than in the suburbs of Reno. We would therefore anticipate that stores would be closer together in Manhattan, that they would be smaller, and that the prices charged would be greater. This anticipation, of course, would be found correct if we looked to the real world. Unfortunately, this set of hypotheses runs in the opposite direction from those we developed earlier. As long as the density of population is not so large as to raise significant congestion problems, we should expect that increasing density would lead to larger units, decreasing prices, and shorter distances between establishments. At some point[10] congestion would begin to raise transportation costs, and from that point on we would have two effects occurring together. Whether prices would go up or down and whether store sizes would increase or be reduced is thus indeterminate. Note, however, that both effects lead to reduced distances between stores. It might be possible to work out a theoretical prediction of the combined effect of congestion and increasing density on price and establishment size, but I have not yet accomplished this. Summing up, the general picture we would expect is shown in Figure LV. Clearly this appears to be what we see in the real world, although we should make statistical tests rather than relying on impressions.

There is a complicating factor which has so far been ignored, but which requires at least some discussion—land rentals. Our ice-cream trucks pay no rent, but if we think of drugstores, they do. The basic rent charged is dependent, not on the profits of the drugstore in a given location (because usually there are a number of equally good locations in any area), but on the number of other types of retail establishment which might be established there. The man thinking of establishing a drugstore will normally dicker with several potential landlords. Each of these landlords, in turn, will be negotiating with not only the man who wants to start a drugstore but also potential cleaning establishments, small stores of various sorts, and perhaps a lunch-counter project. Under the circumstances, the rent the drugstore operator must pay is not a simple function of the value of his location to him. It depends essentially on the value of the location for retail establishments in general, and, hence, is an extrinsic factor.

The importance of this factor is simply that higher rents are

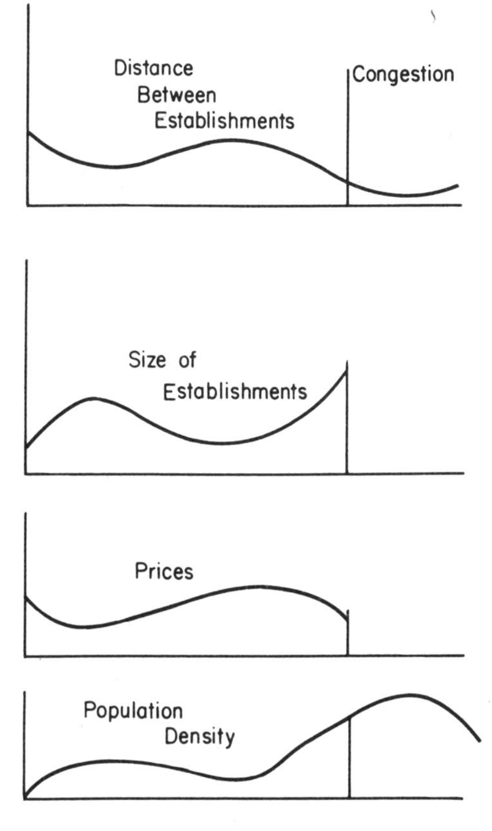

FIGURE LV

essentially a technological change. The most efficient design of a drugstore at a rental of $1.00 per square foot is unlikely to be the most efficient at $7.00. Since we measure size of establishment by volume of business, not by square feet of floor space, it is not at all obvious that this technological factor would make any difference, but it might. In the congested areas where rents are very high, it may be that drugstores require somewhat different combinations of factors and that this leads to some variation from our predictions. Basically, however, this phenomenon would occur only in the congested areas where our predictions are pretty feeble anyway.

So far we have been talking solely about that sort of monopolistic competition which occurs because of the geographical dispersion of the population. The type of monopolistic competition which arises because people have different tastes on nonlocational matters is, I contend, fully analogical to the geographic model. The principal difference is that products can vary in many different ways with the result that the space in which the firms "locate" is multidimensional instead of two-dimensional. For simplicity, however, let us once again confine ourselves to a single dimension. Shirts, for example, can vary in almost an infinite number of ways, but we can temporarily consider only the tightness of weave of the cloth. On Figure LVI we have price on the vertical axis and the weave, from coarse to fine along the horizontal axis. The three irregular inverted U's show how much three customers would be willing to pay for a shirt at various weave finenesses.

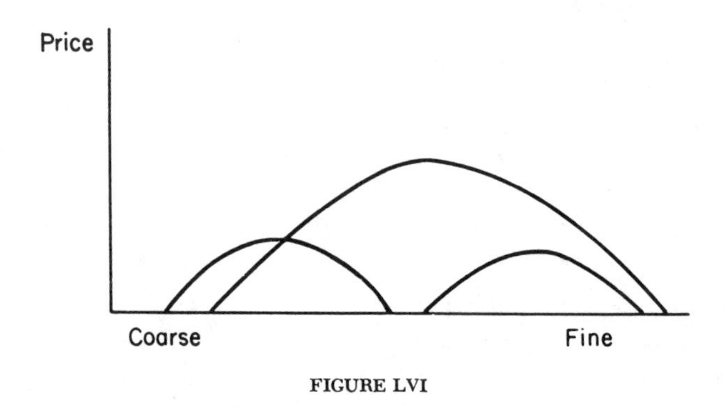

FIGURE LVI

The situation differs from the geographical model only in that I am now using inverted U's instead of V's.[11] Clearly, if the shirt manufacturer knew the location of each customer's optimum and the shape of his inverted U, he could go through the same operation we outlined for the druggist deciding where to locate his drugstorè. If the populace were grouped in some sort of normal distribution, which is what we would expect, then the fineness of the weave of the various manufacturers' production would be related to that curve as shown in Figure LVII.

The manufacturer doesn't have this information, of course, and neither do we, so this might appear an untestable assertion. In fact, however, it is testable with a little elaboration. The manufacturer can find his individual optimum by simply shifting the

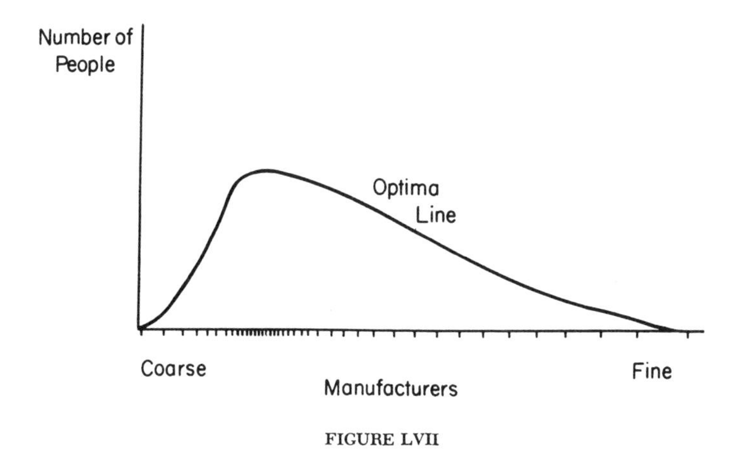

FIGURE LVII

fineness of weave a little and noting whether his profit goes up or down. After a period of time, we would expect the various monopolistically competitive producers to be arranged in the way shown, even if they did not have any direct information on the distribution of the individual optima. The distribution of the manufacturers along this dimension would be observable, and from it we could deduce distribution of the customers. From this distribution, itself untestable, we could deduce that prices would be lower and size of establishment larger in the area where the manufacturers are located most densely, and this would be testable in the real world. Thus, we can "test" monopolistic competition when it is due to product differentiation as well as when it is due to geographical factors.

CHAPTER VI

The Production of Information

One of the major industries of the modern world is the production and dissemination of information and entertainment. Similarly most of us devote a good deal of time to the "consumption" of these commodities. Since we normally pay, directly or indirectly, for the information we consume, there seems no reason why this particular industry should be greatly different from others. The provision of information, however, has special effects which we do not normally expect in other industries. My opinion on various subjects is not entirely a simple product of the various bits of information I have imbibed from the "media," but it clearly is much affected by them. The decision of a magazine publisher that running two articles by Robert Ardrey will increase the magazine's circulation will also have the effect of somewhat affecting the views on the institution of property, held by a very large number of people. A newspaper publisher who helped his sales by providing thorough coverage of the Watts' riots would also promote race tension. This opinion-changing effect of the publication may be an unintended by-product, or it may be desired by the publisher. It is unlikely to figure prominently in the desires of the consumer when he chooses what he will watch on television.[1]

These changes in opinion, however, are of considerable importance in themselves. They have a great effect on the political decisions taken by democratic governments, and they can sharply change the pattern of economic production by changing the tastes of consumers. Whether these effects are "externalities" or not depends on the precise definition given to that term. The purchaser and the seller of some news media are the only ones directly affected if the exchange modifies the opinion of the purchaser; let us say it makes him more tolerant of other races. The fact that the purchaser's opinion has changed, however, may make him take different decisions in future transactions, and in this sense has external effects. If he votes differently, or buys a different product, then the politicians and producers who are affected by these changed decisions feel something which is remarkably like a true externality. Even if it is not really an externality,[2] the matter is of enough importance to repay further study.

This chapter, then, will investigate the business of producing information and entertainment, with the primary concern being the effect of the industry on the opinions of its consumers. Since we are concerned with opinion formation, the whole industry will be treated as a provider of information, not of pure entertainment. Thus, a play which contains something which might change the opinion of some of its audience will be treated as simply an elaborate package for the information. People mainly read, watch plays, movies, and television, and listen to speeches or the radio because they get some direct satisfaction from it, i.e., they are "entertained."[3] Similarly, the producers are primarily interested in attracting purchasers, and, thus, may be relatively uninterested in the information content of their output. Therefore, the information or opinion-changing material is really a by-product, but it is an extremely important by-product.

The normal magazine contains both articles which are purportedly aimed at informing the reader, and fiction which is allegedly intended to amuse him. In fact, however, he would not read the articles if they did not please him, and the stories may have as much, or more, effect on his opinions as the articles. The editors are aware of the first fact, and both the subject matter and the treatment of their articles are clearly designed to attract readers. The opinion effects of the stories may be given less conscious consideration, but most magazines give at least some attention to the matter. Normally, this attention results in little more than a policy of avoiding certain themes which, it is thought, will annoy the readers. In most cases, however, getting readers is the great objective of the magazine, and the information content of the magazine is shaped to that end.

Magazines, of course, vary greatly in the information they contain. At one end, *Popular Science* is devoted almost entirely to factual information. Its selection and treatment, however, is clearly intended to attract the readers, and the editors will frequently "sensationalize" some bit of information about the advances of science almost completely out of recognition. At the other extreme, there are magazines devoted exclusively to fiction. The other media show a similar spectrum, from factual information to fiction, and similarly are basically interested in attracting consumers, with "opinion formation" simply a by-product. If we turn to media which appeal to the intellectuals, we find the same picture, albeit at a higher level of intellectual difficulty. Perhaps the intellectuals are more readily influenced by what they read because many of them feel it important to be "in," and hence try to guess what the next fad will be before it arrives. They want to have the latest ideas and tastes, and this desire makes them much more eager to adopt new ideas, right or wrong.

For our purposes, however, we will ignore those aspects of the "entertainment" function of media which do not involve information at all. A play will, presumably, have a good deal of noninformation content, but we will leave this to other scholars. For us only the information that it contains is interesting. Note, however, that this does not mean that we will ignore all of those aspects of the play which might attract an audience. To some (perhaps very small) extent the play will contain information about the world in general, and this information will be one of the factors which will lead it to be a success or a failure. Shakespeare, for example, did a great deal of propagandizing for the crown and, in the course of this propaganda, put a lot of "history" into some of his plays. This not only kept him on excellent terms with the censors but it also no doubt added to the box office appeal of his plays to the patriotic yeomanry. The more highly educated classes, who knew how he distorted the record, might, on the other hand, have found this "information" a defect in his plays. Still, on the whole, he no doubt increased the box office by his rather unfair presentation of "the facts." Mostly, of course, his audience appeal was derived from other factors, but surely the information content of his plays had at least some effect.

Discussing Shakespeare solely as a purveyor of information may seem bizarre, but confining ourselves to this aspect of his plays will greatly simplify our analysis. A similar simplification of the output of less exalted "media" will also make it possible to develop a theory of information flows which includes this material. In fact, most information is transmitted mainly as a by-product of entertainment.[4] Thus, our theory will be more helpful than if it confined itself to information transmitted for its nonentertaining characteristics. But this requires a warning. We will talk mainly about information, but the "wrapping" of pure entertainment may in many cases be vastly more important in the minds of both the producer and consumer. But the fact that information may be only a minor part of the "product," does not prevent our discussing it. In many cases, the newspaper will do as an example, the information is the heart of the matter. Even here, however, the information is provided mainly to interest the reader. It competes for space with the funnies and the astrology column in terms of how many readers it will attract, not in terms of its intrinsic merit.

The problem may be treated in a fairly rigorous manner through the use of models which have a great formal resemblance to those used in the last chapter. The total volume of information[5] is a multidimensional complex, but we can confine ourselves to variations along one dimension in the same way as we can consider the many ways in which a product may differ one at a time. Suppose that one of these dimensions is shown by the horizontal line of Figure LVIII. Individual A's present state of information is shown by his location.

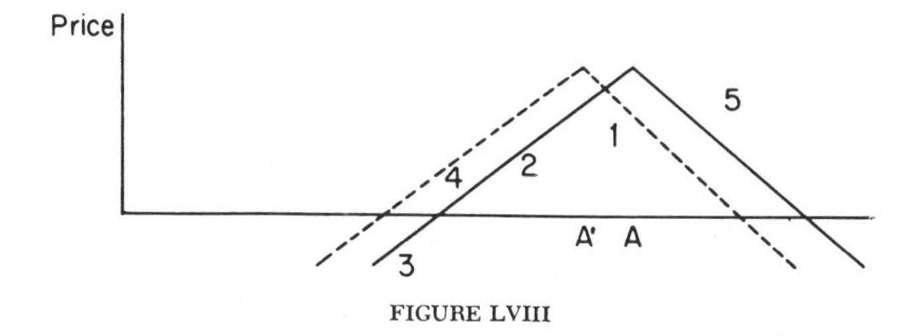

FIGURE LVIII

Let us further assume that A is a very stodgy individual. The type of information he most wants is exactly the information he now holds. He wants his opinions reinforced, not changed. This may seem a most unlikely assumption, but we will stick with it for awhile and develop a completely non-dynamic model of a world in which nobody wants to learn anything new. Unlikely as it may seem, this model will tell us something about the more complex real world. It will also serve as an introduction to more realistic models. A, then, would buy exemplars of media at prices shown by the "tent" in Figure LVIII in much the same way as our customers of monopolistically competitive manufacturers. In fact, of course, he is a customer of monopolistically competitive manufacturers of media.[6] We shall change the assumptions of the last chapter a little by permitting our customers to buy all media which fall, in terms of content and price, within their "tent" instead of simply the one which fits their needs best.

Customer A, then, would buy magazine 1 at a fairly high price and magazine 2 at a lower price. If the operators of 2 raised its price significantly, he would stop buying it because it is farther from his present state of information than is magazine 1. Magazine 3 illustrates a theoretical point which may be of some interest. Customer A would consume the information in 3 only if he were paid to do so. It might well be that a magazine which is very attractive in its noninformation aspect could be said to sell its information to some of its customers at a negative price. I have sometimes found people subscribing to magazines which they regard as highly unreliable, because they like the pictures, fiction, etc. In a sense, they are repelled by the "information" but receive payment in other matters, the packaging, and take the journal. This theoretical possibility, however, will not be further discussed. From now on we will assume that the customer always pays a positive price for his information.

Granted that customer A is consuming magazines 1 and 2, he will gradually find the information he holds, changing. This is simply because magazine 1 gives him exactly what he now believes and magazine 2 gives him some different information. This will shift his state of information, over a period of time, to the left, say, to A′. At A′ he will also begin to read magazine 4, which will lead to still further leftward shifts. Thus, our stodgy reader, who wants only to have repeated what he already knows, finds his opinions affected by what he reads. Since what he knows is a function of what he has read, and since the market provides him with information which is not identical with what he already knows, his tastes are changed. The process would lead to a random walk by the customer until he found himself in a position where his reading no longer changed his position. Our customer is at such a position at A′, but this merely reflects the spacing of the magazines on this particular diagram. With another arrangement of media, A might find himself in motion for a very long time.

One aspect of this model is of considerable importance for our further analysis. Magazine 5 is much closer to A than is magazine 2. Nevertheless, the lower price of 2 means that A will read it and not 5. Thus, the customer is moved to the left by 2 and not moved to the right by 5 because of their different prices. In general, the lower the price of the information in a given media, the more it will be able to attract consumers who are far removed from its position.[7] Thus, the lower-priced journals are able to pull customers toward their position, while the more expensive sources of information have much less attraction power.

Let us now consider the situation in which there are many potential consumers of information. At the top of Figure LIX we show a population which is assumed to be distributed in the conventional bell-shaped form on one of our opinion dimensions.[8] We assume that all of the customers are like A, that they will pay a higher price to be told what they already believe than anything else. We also assume that the producers of "information" are simple profit-maximizers, who try to get the best return on their money by proper choice of their editorial policy and price. The editorial policy takes the form of selecting some point on the issue dimension, and we shall assume the same conditions of monopolistic competition as in the previous chapter.

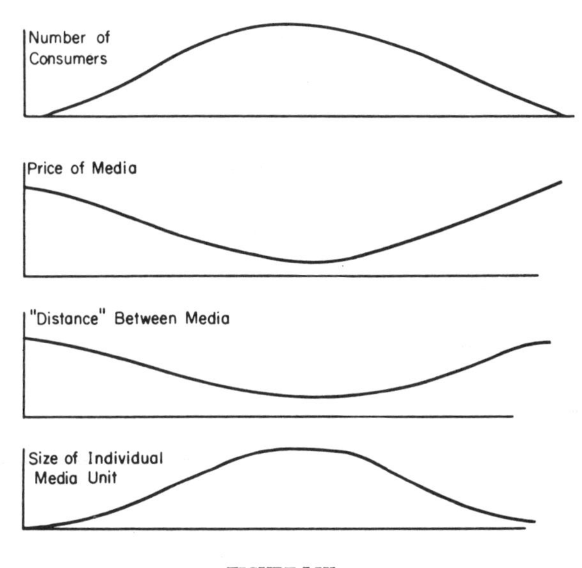

FIGURE LIX

This makes it possible to directly transfer to our present case some of the results of the last chapter's reasoning. The second level of Figure LIX shows the price charged by media in various positions, the lowest level being in the areas where customers are most concentrated.[9] Similarly, the third level shows that the media's editorial policies will be closer together in the densely populated area and that they will be larger, as shown in the bottom level. These responses of the producers to the distribution of the customers, however, result in further changes in the customer's distribution. As we pointed out earlier, lower priced information sources tend to move individuals toward their position. Thus, the bell-shaped distribution of opinion among the customers will become more peaked, and this in turn will strengthen the factors shown in the three lower levels, which in turn will lead to still further peakedness in the customer distribution. This process will continue until an equilibrium is reached.

The tendency of the interaction of the original distribution of opinion and the organization of the information-producing industry to produce a concentration of opinion is further strengthened by the fact that not only are prices lower near the center of the distribution but the distance between individual media is smaller. In Figure LX we show an individual, who is located to the right of the center of the distribution, and the media close to his position. Note that the media become more expensive and less frequent as you move to the right. In this particular case the diagram has been drawn in such a way that the different prices charged by the magazines have no effect. The fact that the distances between magazines to the left of A's present position are less than between those to his right, however, will result in his reading more magazines on the left side, and hence will eventually shift him in that direction. Altogether, the tendency of the information-providing industry to "standardize" opinion is a strong one.

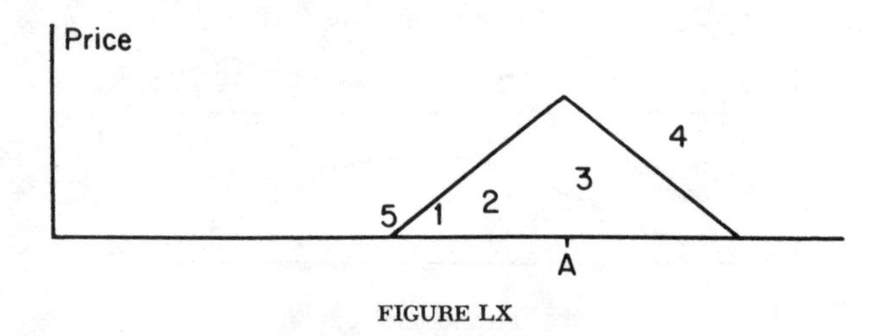

FIGURE LX

To a considerable extent, the strength of the phenomena shown in Figure LX depends upon the size of the economies of scale in the industry. It happens that with present-day technology, they are very great. Think how much a copy of *Life* would cost if only 50,000 were sold. Thus, we would expect the concentration to be much greater than if the economies of scale were as modest as they were in the Middle Ages. The television industry, probably our primary opinion-former, is an extreme case of concentration, having only three networks.[10] Presumably, this is the result of the policy of the FCC rather than economies of scale, but we can still analyze the results with our present model.

There are two special features of present-day television in the United States: it is free to the customers and only one network at a time can be tuned in on your set. The fixed price of zero means that it is impossible to design programs for a small number of customers who are willing to pay high prices.[11] Thus, there will be an even greater tendency for the networks to concentrate on the bulge in the opinion distribution than for other media to do so. This, again, will lead to further concentration of opinion. The fact that customers are forced to make a clear choice of networks may well have a similar effect. In Figure LXI the vertical axis shows the consumer surplus which a consumer at some given location gets from tuning in to each of three networks, A, B, and C.

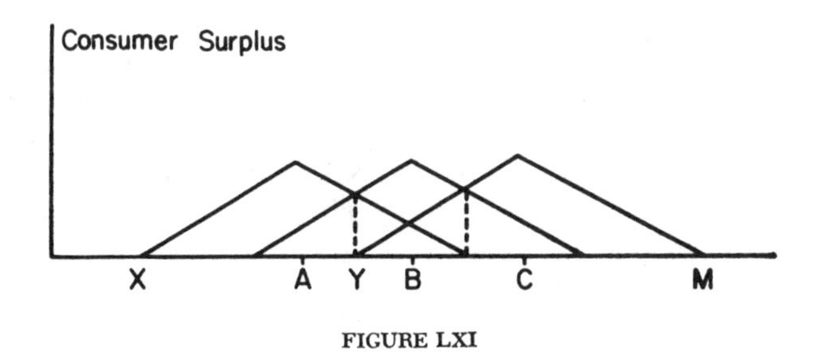

FIGURE LXI

The construction can be easily derived from the tent-like preference structure of the individual preferences which we are using. If the consumer surplus is positive for some station, the consumer will turn on his set. If it is positive for two networks he will choose the one with the highest consumer surplus. Thus, consumers to the left of X and to the right of M will leave their sets dark. Those between X and Y will watch station A, those between Y and Z will watch B, and those between Z and M will watch C. Note, however, that this tends to move the extreme opinions toward the center. The distance AY is necessarily shorter than the distance AX.[12] Thus, the station's information output is not in the center of the opinion spectrum of its watchers, but nearer the center of the total spectrum of all persons. This may further tend to pull the watchers toward the center.[13]

The net effect of our discussion so far is that the interaction between a bell-shaped, initial distribution of individual opinions and a set of profit-maximizing producers of information, under conditions of monopolistic competition, will lead to further concentration of opinion near the center. This model, however, is static, and has no potential for any movement after its stable equilibrium is reached. Further, it depends upon a most conservative assumption about the motivations of the individuals in society—they are assumed to want to consume more information just like the information they already have. If we change this assumption, we may be able to get a more realistic picture of the world. The first change to be examined will produce an output much like our present model, except that the whole structure is engaged in a random walk instead of being stable.

Marshall McLuhan was once told by an editor of McGraw-Hill that: "A successful book cannot venture to be more than 10% new."[14] Dr. Norman Storer made somewhat the same point in his *The Social System of Science*.[15] Let us, therefore, assume that our information consumer wants some new information, but not too much. That he is interested in novelty, but doesn't want to get too far from familiar territory. The simplest way to incorporate this assumption into our model is to assume that his preferences are shaped like the solid line in Figure LXII. He prefers a small shift in information, because of its novelty effect, to a repetition of his present information. He remains conservative enough so that large changes make him unhappy.[16]

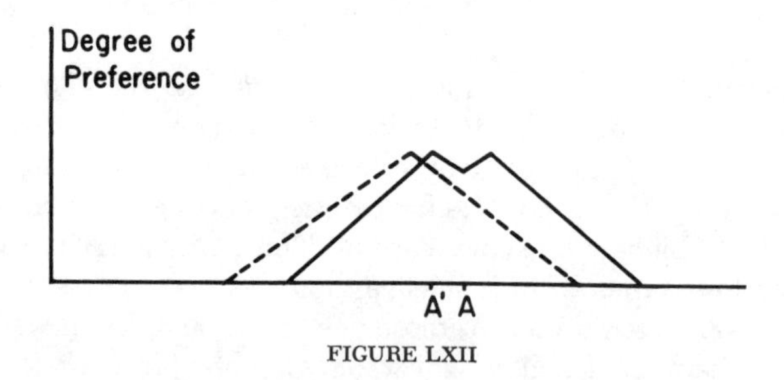

FIGURE LXII

Once the individual has received new information, his position shifts to some point like A' in Figure LXII. Here again he will prefer information which has some element of novelty rather than simply a repetition of what he already knows. In this case, however, his desire will be asymmetric. Information which is similar to what he held before he moved will not appear to him to be novel. Consequently, his preference for information would take on the form of the dotted line in the figure. In practice, however, this does not mean that an individual will simply be moved steadily in one direction if information exactly meeting his desires is provided. In our figures we have only one dimension out of very many, and novelty can be sought down any one of them. The individual might engage in a random walk in this multidimensional information space. The only restriction would be that he never double back. There is no necessary tendency, however, for him to continually move in one direction on any given dimension.

The problem which a universe of consumers, who have this type of a desire for variety, presents the producers of information is a difficult one. Each day's output must contain an element of novelty. In our previous model, the publishers, TV directors, and playwrights could follow the rule which characterizes most of our economy, and continue to produce exactly the same product for considerable periods of time. Every issue of *Life*, on the other hand, must be completely different in editorial content from any other issue, while at the same time keeping somewhat the same basic position. The one thing the editors know is that they cannot simply continue to produce an issue which seems to be unusually successful. Further, they can hardly collect information on what the people want, because people can hardly tell you what information they would like to hear if the information which will attract them will do so because it is novel. The novel aspects have to be new to them, and therefore they cannot specify them in advance.[17] Theoretically, it would be possible to experiment with a number of different products and a number of separate samples of consumers, but it would be extremely expensive.

It would, of course, be possible for the various random walk patterns of the individual consumers of information to cancel out to an approximately stable pattern of the whole body of consumers, so that the media would not have to continually change their positions. Observationally, this is not what happens. Obviously, public opinion is not a stable, constant phenomenon. There will, for example, suddenly be great concern about automobile junkyards, then, after a while, the subject will sink back to the obscurity from which it came. The focus of the system of public information is continually shifting.

This continual shifting of the desires of the consumers is one of the reasons why the industry of producing information is so much more uncertain than that of producing automobiles. The strains of magazine publishing are greater than those of, say, running a railroad. This is reflected in the much more conspicuous role of emotional problems in this particular industry. Artistic temperament may be an inherent characteristic of the "creative" personality, but it may also merely reflect the intellectual strain of trying to guess what the public, in a highly unstable environment, will want next. The reasons for this instability of the public desires, the explanation for the fact that the individual random walks of the consumers do not produce a stable total configuration will be discussed later. Let us, for the moment, simply accept that this is so and examine its consequences for the producers of information.

The publishers and editors of the various media know that what their audience wants is continually changing, and that it is important to anticipate these changes. They should, therefore, be continually seeking information on what will be a salable variation. An obvious method to use in this search is simply to see what the other media are carrying, and note the success that they have had. If *Life* runs a picture feature on automobile junk yards, and it seems to be successful, then *Look* would be well advised to do the same. In the first place, presumably they don't have exactly the same readers, so for some of their readers it will have the feature of novelty. Further, they have evidence that this particular type of novelty appeals to the "public" at this particular point in time. In a sense, it is a certified "seller" and it can be presented with enough difference from the *Life* article so that it will still have an element of novelty even to those readers of *Look* who have read *Life*'s feature.

Thus, with each media editor trying to find a subject and treatment which will help the circulation of his output, there will probably be a great deal of "follow the leader" type of activity among the media. In the absence of other information as to what will "sell," the editors simply note what seems to be successful in other magazines. This means that the typical history of some particular treatment of some subject is typically tried out in one magazine, repeated with variations in several, spread to almost every media, and quietly abandoned as it becomes "old hat." There is even a regular division of labor among media. Certain limited-circulation magazines try always to have new ideas. Some of these catch on, and are then tried by more general media. Thus, Camp started in the "little" magazines, and gradually spread until Batman was put on nationwide TV.

With the sources of information open to the individual tending to move in a coordinated manner, an individual who wants the information which he will consume to be like that which he already has, but with some element of novelty, is likely to find that his position in the multidimensional information space of our model moving in much the same way as that of his fellows. Not the individual but the group will be following a random walk. As a result, the bell-shaped distribution itself moves randomly over a fairly large area. This pattern could be deduced simply from the behavior of the media, as we have done, or it could be due to some other, more basic, phenomenon. Perhaps people like to have somewhat the same opinions as their neighbors, like to "be in fashion," etc.

If it is assumed that it is the tendency of the media to follow each other because they lack other information as to what will sell, this will lead to a pattern of opinion in the public which makes continuation of this policy profitable. In a sense, we have a circular chain of reasoning, with the behavior of the media causing behavior on the part of the consumers, which makes the media behavior rational. Such reasoning is always somewhat suspect, but there is no need to reject it out of hand. There are many phenomena which must be initiated by accident or some other factor, but which are self-sustaining once started.

This model provides for movement of public opinion, but gives that movement no direction. A random walk is the predicted shape of the changes which will be brought about. This model, in my opinion, does describe the behavior of opinion in many fields, particularly those where fashion is an important factor. How long dresses will be this year, which painters (including painters of long ago) will be "in" this year, matters of taste in general, seem to follow just the sort of random movement we would predict. There is even a distinct tendency for movement in some given direction to be continued for a while, which follows from our model. Direct reversals are most unusual, even in skirt lengths. Normally skirts get shorter for several years, then pause for a while as other changes are made in dresses, and then lengthen again.

Still, not all changes in opinion have this random pattern. We hope, certainly, that in some areas we are making progress, and progress would be a nonrandom element of direction. It is also obvious that there are people who are able to influence opinion in directions which they favor, and this also is a nonrandom component.[18] Let us try to complicate our model to take these phenomena into account. First, we have so far assumed that the individual's preferences for information depends entirely on the information he now has. Surely it is normally so, that an individual also has preferences as to what the information is. If I am awaiting a medical test to tell me whether or not I have cancer, either answer might be equally distant from my present state of uncertainty, but they will not make me equally happy. We need a model in which the individual's interest in acquiring information is dependent both on the "newness" of the information and the inherent interest of the information.

Algebraically, the problem would be simply solved by some equation of the form:

$$A = f(D, I)$$

A = attractiveness of information

D = degree of difference from present information

I = inherent attractiveness of information

Since we have been using geometry in our work so far, however, a geometric model would be more fitting here. We can modify our existing model by introducing a geometric representation of the superior inherent desirability of some particular information. In Figure LXIII we show the inherent attractiveness of various positions by the height of the shaded area above the axis, and then geometrically add the same preference structure for information used in Figure LXII to get a measure of the total attractiveness of each information complex. B is assumed to be inherently more attractive than any other point on the horizontal axis, the individual now at point A has the same desire for some new and some old information, as in Figure LXII. The total curve, however, is quite different.

Note that individual A does not now realize that he would

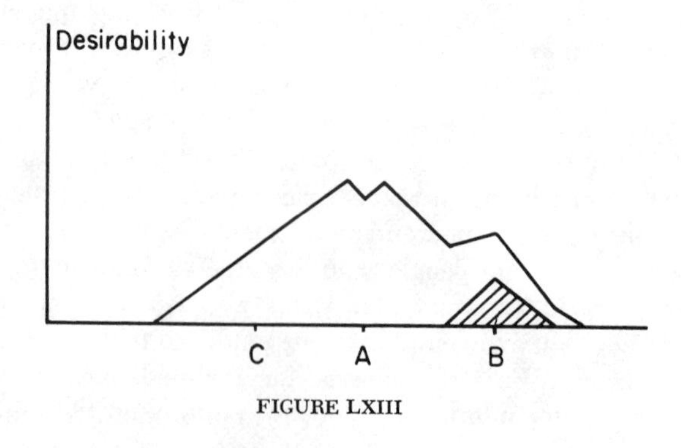

FIGURE LXIII

be attracted by the information represented by B. If he did, he would already be there. The figure shows the attractiveness of different bits of information looked at from the outside. We assume that if an article appears, giving information B, and another, giving information C, he will get more pleasure out of B, and, hence, will be more likely to continue reading that magazine. The possibility that he was attracted to B by simply having it brought to his attention by the title of the article on the cover of the magazine should not, of course, be overlooked. In this case, however, the information situation before he sees the magazine cover is that shown in Figure LXIII. After he sees the title of the article, he is informed and his attraction to the subject of the article would move him to the right. The producers of information must make guesses as to what will attract their readers, and these guesses may well be wrong.

As can be seen from the diagram, even if the inherent attractiveness of some bit of information is considerably smaller than the individual's attraction to information because of its position, it exerts quite an influence on the type of information which it would be wise for the media to offer. Further, if the inherent attractiveness of some given position is "interpersonal," i.e., if it is inherently attractive to many persons, then it is likely that all of the media will move to that position.[19] One would anticipate that, after a whole, the media would be distributed as in Figure LXIV. The reason that the media would not be at absolutely the same position is simply that some people have very strong preferences for variety, and these might outweigh the inherent attractiveness of B for a small part of the audience.

One example which may be used to illustrate the phenomenon is the progress of science. A new discovery, if it is accompanied with sufficient evidence, immediately becomes inherently very attractive to the students in its field. After it has been appropriately tested, they will all move to that point, as will the scientific

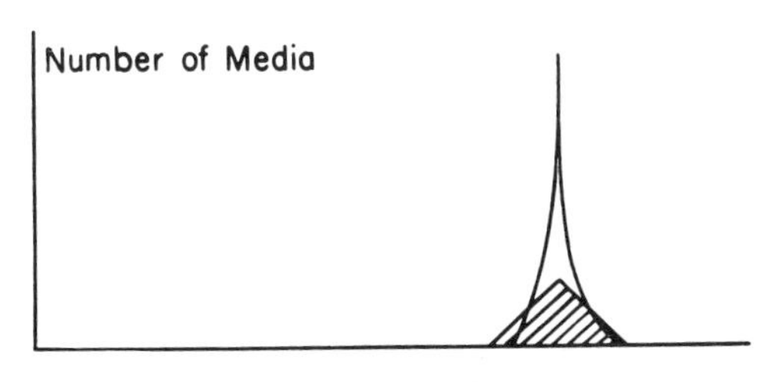

FIGURE LXIV

journals.[20] Since novelty in science is readily available from other new ideas, the result would be a straight vertical line, instead of the narrow bell of Figure LXIV. The scientist, in fact, is much like any other consumer of information. The principal difference between science and ordinary information exchange is simply that science is organized to make the recognition of truth easier, and, hence, that inherently desirable information is more common. When scientists deal with a problem on which the evidence is weak to nonexistent, they are as likely to follow a random walk as is the ladies fashion industry. The cause of cancer, for example, is unknown, but for the last fifty years researchers in this field have tended to swing from one potential explanation to another in much the same way as dress designers change the height of skirts.

It should be noted that even in matters of fashion, the evidence indicates that inherently superior points may exist. The most obvious examples concern technological progress. Improved materials make changes in design possible, and these changes may be made by everyone and then persist over time. The changes associated with the introduction of "stretch" fabric will do as an example. Even when there are no technological changes, apparent points of superiority may develop. In the nineteenth century, men's clothing moved into a relatively standardized pattern which has persisted ever since (it may now be changing to another pattern). Not that all men dress exactly alike, but the range of variation is small compared with the range of women's dress or the range of men's clothing before 1800. Presumably, this reflects merely a particularly strong "fashion" shift, but there is no theoretical reason why we might not have inborn preferences for certain types of clothes, foods, etc. Discovery of such inherent superiorities would be worth a sizable fortune to people in the industry.

So much for the differential inherent attractiveness of different sets of opinions. We now turn to the manipulation of opinion by public relations techniques. It should be noted that we will confine ourselves to such manipulations in relatively free markets for ideas. Conversion of our model to handle totalitarian propaganda would be fairly easy, but is beyond the scope of this study. Our first example of opinion manipulation will be a very simple one, an advertisement. Clearly, most journals would get more readers, and make them happier, if they filled their pages with items intended to attract the readers, but instead they sell the right to put in items which will have very little attraction to subscribers. If we think of the editor and the advertiser as being combined in a joint enterprise, they sacrifice circulation in order to sell an idea (Buy Buick) which they favor.[21]

If the management of the media wishes to push some idea, whether because they are devoted to it or because someone will give them some quid pro quo, they must accept somewhat smaller circulation in return.[22] The profit-maximizing executive will not try to use his journal to express his own point of view. In practice, the major newspaper chains exemplify this attitude, with different newspapers in the chain having different editorial policies. But no one, of course, would contend that all media are controlled entirely by profit-maximizers. In the first place there are journals set up specifically to push some point of view. The *New Times* and *America* are well-financed examples, but at any given time there will be a number of fringe journals put out by private persons and clearly not attempting to maximize their readership. Sometimes they become commercial successes, as the *National Review* has, but the history of the *Nation* is more typical. Even in the case of the *National Review*, it seems that the profitability of the journal is an accidental by-product of its excellent editing rather than of its political position.

In part, then, the media are controlled by people who are not solely interested in profit, and it must be assumed that this has some effect on public opinion.[23] People who read these journals have a somewhat different information demand than people who do not, and the unwillingness of the journal editors to shift their position in order to conform to the information demands of their readers means that the whole structure of information is moved a little. In Figure LXV the density of population holding each view along a portion of an information dimension is shown, and the "profit-maximizing" position for a set of journals selling to them by points A, B, and C. Suppose that the owners of journal B have strong views and shift to point B′ in spite of the probable decline in profits they will experience. This not only changes their position, it also moves the profit-maximizing position of journal A to A′ and C to C′. Thus they have made a sizable shift in the information available in this part of the spectrum. Over time, this should shift the information position, and, hence, the information demand of the customers from the solid line to the dotted one. The owners will not make as much money as if they had stuck to B, but they may regard this as unimportant.

Similarly, even if the owners of some information media are uninterested in anything but profit, their employees may have axes

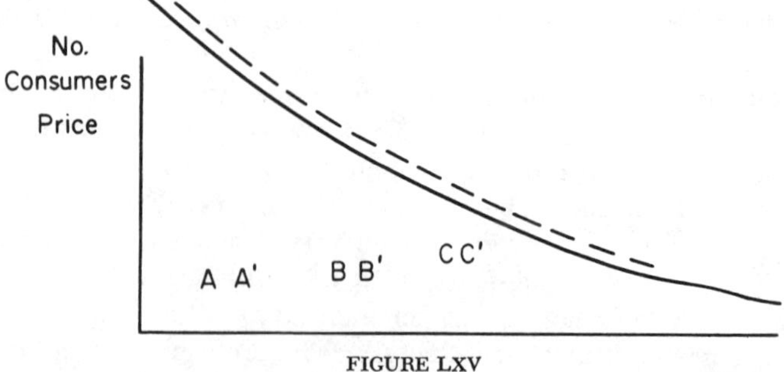

FIGURE LXV

to grind. Here, again, the employee solely interested in maximizing his income would simply try to please his employers by getting the most readers, but most people are not that simple, and some efforts to maximize utility at the expense of income can be expected. We will return to this subject in the next chapter. It should be noted that the exceptional problems faced by entrepreneurs in this field in making estimates of the wants of their customers, make it particularly difficult to determine whether a profit-maximizing mix of information is being presented. Thus, a trusted editor might "get away" with much more in the way of imposing his preferences on the customers than would be possible in other industries. Similarly, both the editor and the entrepreneur may be misled by this lack of clear information and think that what they prefer for personal reasons is also the optimum information pattern of their customers. All of these factors may move the journal away from the optimizing position in terms of profit, and have their effect on public opinion indirectly.

Implicit cartels may also develop in this field. Suppose, for one reason or another, most of the owners of media or most of their subordinates develop the same general attitude on some given subject. Then the customer who is dissatisfied may find nowhere to turn, and the profits of the group of journals will be the same as before. Over time, this concentration of all the media on one point of view might shift public opinion to a structure in which the consensus of the media becomes the consensus of public opinion. Under these circumstances (and I have no idea how common they are), it might be quite hard to get an opinion with very great potential appeal even before the public. The difficulties faced by the founders of the *National Review* may serve as an example. They were subjected to what amounted to a boycott by the rest of the press, which made their growth extremely difficult even though their eventual success proves that a suitable market existed. Only the great wealth of the Buckley family made it possible for them to continue and eventually break the cartel barrier.

These methods of influencing public opinion through the media may be matched by somewhat similar methods by the customers. An editor can influence the information received by the public if he is willing to sacrifice a bit of his prospective profits. Similarly, a consumer can influence the structure of media, and hence the shape of public opinion, if he is willing to sacrifice to that end some of the pleasure he gets out of consuming information optimally. His consumption not only influences his own state of knowledge, it "signals" the producers of information as to what will sell and make money. If I were to cancel my subscription to a journal, although subscribing is in net to my taste (I could continue to read it in the library), the effect on that journal and its competitors would be the same as if I had dropped it because I was dissatisfied. Naturally, the effect of one customer's action would be *very* small. Thus, to some extent, I would misinform the producers of information, and they would act on the misinformation, resulting in a change in the structure of publications and of opinion. Suppose, for example, that people at the left of the opinion spectrum were to decide that they would no longer read *Time* until it shifted its editorial policy in their direction. If this were not countered by any other influence,[24] it would shift the general structure of media to the left, which in turn would have an effect on general public opinion, which the leftists would find desirable.

It is, of course, easy to see this phenomenon as more important than it is. I know of no case where this strategy has been successfully applied, but it is theoretically possible. Another, and probably more sensible, technique would be to write letters to the editors of journals. As has been noted, the producers of media live in an environment of restricted information. Under the circumstances, they may put disproportionate emphasis on the opinions expressed in the rather few letters[25] which they receive. It would, therefore, be fairly easy to organize a systematic deception of them and, hence, a distortion of public opinion. If every time some sort of article were to appear, ten to twelve people would write in saying that they think it is wonderful (or terrible), this might give the editors a completely erroneous idea of what their readers want. The limits on this technique are, I suppose, obvious enough, but that it can sometimes be useful to persons wishing to change public opinion (or prevent such changes) should be equally obvious.

CHAPTER VII

Political Ignorance

Having discussed the provision of information by the media, we will, in this chapter, cross the street and deal with the individual "consumers" of information. Primarily our concern will be with information costs and their effect on political decisions. Any discussion of this subject must necessarily owe a considerable debt to Anthony Downs's *An Economic Analysis of Democracy.*[1] Although I will follow along the general route he pioneered, the technical details will generally be different from the Downs's model.

We may well start our consideration of information and politics by taking seriously a well-known joke. Mr. Smith, upon being asked who made the decisions in his family replied, "We have a division of labor. My wife makes decisions on minor matters and I make them on major problems. For example, my wife decides where we should go for vacations, the children's education, etc. I decide our attitude toward the recognition of China." It is always a mistake to analyze a joke, but let us, at least, inquire what is funny about this one. Most people would agree that relations with China are, indeed, more important than where the Smith family spends its vacation. However, most people would also feel that, in fact, the wife runs this family. This apparent paradox is easily explained. The decision of the Smith family on where they will spend their vacation will be the controlling decision on that matter, but although in a democracy their decision on relations with China has some influence on the country's relations with China, it is clearly only a tiny amount. If we evaluate the importance of the minor influence which the attitude of the Smith family has on our relations with China, then we will see that the Smith family decision on China is, indeed, much less important—as regards effects—than their decision as to where they will spend their vacation.

Public problems are normally more important than private problems, but the decision by any individual on a private problem is likely to be more important than his decision on a public problem, simply because most people are not so situated that their decision on public matters makes very much difference. It is rational, therefore, for the average family to put a great deal more thought and investigation into a decision such as what car to buy than into a decision on voting for President. As far as we can tell, families, in fact, act quite rationally in this matter, and the average family devotes almost no time to becoming informed on political matters, but will carefully consider the alternatives if they are buying a car.

The immediate reaction of the reader may be that this description of the behavior of the voting and buying public is not that which he observes among his friends and acquaintances. He may, of course, be correct in his observations, but they are made in a nonrandom sample. The reader is probably much better informed and more interested in political matters than the average. Since like appeals to like, it is probable this his friends are also much interested in political matters. Thus, he can hardly judge the average man on the basis of his friends and acquaintances. Not enough research has been done on the amount of effort put into political study by the common man, but it is reasonably certain that the figure would be small.

The individual may, of course, get a good deal of entertainment out of observing politics. He is apt to root for one team, just as he would if he were a baseball fan, and be interested in obtaining information about his heroes. As anyone who has had any contact with baseball fans knows, they do tend to acquire quite a bit of information about the team they favor and also about its potential opponents. This information, however, seldom causes them to shift their support from one team to another. The purpose of the information is to permit them to carry on the enjoyable occupation of conversing about their team, not as a basis for decision as to what team they will favor this year. Similarly, most people acquire political information so that they can participate in the conversation at parties, not in order to decide how to vote. Since strong partisanship is an asset in such conversations, a relatively objective interest in the facts is not likely to be prominent.

Both baseball fans and people interested in politics as a hobby find it easy to get a good deal of information on these subjects because the news media give them good coverage. It would be interesting to find out how many people read each sort of news in the average newspaper, but surely almost everyone, at least, sees the principal headlines and thus comes to know various things about both politics and baseball. The difference between the intent hobbyist and everyone else is simply that the intent hobbyist reads the whole article in areas of his interest instead of merely glancing at the headlines. Even the intent hobbyist, however, will normally depend solely upon news media for his information about politics. The widely sold books of statistics of baseball have no counterpart in the political world (although the revival of political pamphleteering, which was a feature of the 1964 campaign, may change this). Given the low importance of the decision on how to vote, it is rational to depend upon information which comes in automatically through some regularly available source rather than engaging in serious research. Clearly, most voters have always acted this way in spite of the horror of political scientists before Downs.

But if the rational voter is ill informed, it still is true that voters, at least some of them, change their political position from election to election. It is these voters who do change their position who are most important to the political parties, not those who stick to one party regardless. Further, at some elections these shifts can be very large. Lastly, the United States has the primary system, and every man running for office must first win the nomination in a little election confined to the voters of his own party. Clearly, in these primary elections the firm commitment which most voters have to a single party is not relevant. In the primaries not only are the voters not guided by partisanship, they also have even less information than in the regular elections. The newspapers and other opinion media give much less attention to the candidates for nomination, since there are more of them. This means that the amount of information readily available is very small. Thus, it is highly probable that the primary voter is even more ill-informed than the voter in the regular election.[2] The ultimate outcome of our reasoning is that the basic decisions as to who shall hold public office are made by rather poorly informed voters.[3]

In order to discuss the matter with greater rigor, I would like to make use of several rather simple models. The first of these, the "rational-ignorance" model, assumes that the individual, in considering any political issue, makes a sort of estimate of the likely effect upon him of the ultimate governmental decision and of the likely cost of getting enough information on the problem so that he can understand it and take appropriate action.[4] If the estimated effect is less than the cost of becoming informed, then he will not bother with the matter. Suppose, for example, the individual feels that the likely effect on him of the Kennedy round of tariff cuts will be less than $500.00, but that the investment in time and effort necessary to really understand the economic issues involved would be greater than $1,000.00.[5] Clearly, he would be sensible not to bother with it.

It might be objected that the individual cannot make estimates of this sort without carrying out an extensive investigation. In practice, I think that people do use some sort of rough procedure like this in deciding about what they will worry, but the reader need not agree with me. In the next chapter we will shift to the "casual-information" model, which avoids this problem. Meanwhile, let us work out the implications of rational ignorance. The first thing to note is that the politician, in making up programs to appeal to rationally ignorant voters would be attracted by fairly complex programs which have a concentrated beneficial effect on a small group of voters and a highly dispersed injurious effect on a large group of voters.

Note that at least some complexity is necessary. A proposal to tax everybody one dollar in order to give 1 per cent of the population a gift of $100.00 apiece would not work since the cost to the 99 percent of the population in understanding the purposes and effect on them of the legislation would be less than $1.00 in such a simple situation.[6] If, however, the politician can work out a complex arrangement for doing the same thing in a less clear way, he may have a winning issue. Suppose, for example, a program is proposed to strengthen the national defense by keeping the glove-making industry in existence, it being alleged that the glove-making factories are readily convertible into factories for producing some special type of military equipment. If it would take the average voter the equivalent of $10.00 to find out whether or not there will be a significant benefit to the national defense, and, if the likely effect on him, cost or benefit, is less than $1.00, clearly he would be rational in ignoring the issue. The small minority of glove-makers, on the other hand, would be completely irrational if they did not invest the $10.00 in obtaining the necessary information. Thus, the simple fact that the program has the "right" degree of complexity means that the politician proposing it can feel fairly safe in assuming that the people who will gain by the program will know about it and, thus, have their vote affected by it, while those who will be injured will not.

It may, of course, be possible to introduce a completely artificial degree of complexity. Farm subsidies are a clear and obvious example of an income transfer from the majority to a minority. The advocates of the program, however, have succeeded in so muddying the water that the average voter would have to undertake quite a program of study to appreciate its simplicity. Almost any readily available source of information on the farm program will present, in addition to the true picture, a set of rather complex rationalizations of the program. This makes what is basically a simple matter quite complex to the average voter. This artificially introduced complexity no doubt is one of the major reasons for the fact that this fairly straightforward transfer of income has been not only adopted but has grown in magnitude as the number of farms shrank.

One of the major themes of the "Chicago School" of economics has been the desirability of relatively simple and straightforward policies for such problems as cycle control and income equalization. It may well be that their simplicity is one of the reasons they have not been adopted. The people who would be injured by the change from the status quo can easily figure this fact out. Frequently, also, these Chicago School programs will inflict a sizable injury on some small group which is now receiving some sort of special treatment while giving a small benefit to a large group of people. If our present assumptions were to be accepted, it would normally be the case that the small group would take the trouble to understand the issue, while the majority, if there were any significant degree of complexity, would not. Hence the proposal would fail to attract votes.

It is unfortunate, also, but some political issues seem to be such that the information problem is exactly the reverse of what might be hoped. A program like TVA may appear quite complex to the many voters scattered throughout the United States who are injured by it. Since their individual injury is small, they are wise not to make any effort to understand the issue. The people living in the TVA area, on the other hand, are not only the recipients of sizable benefits but appreciation of these benefits is a very simple matter. It is easy to see that the electricity bill is smaller. Thus, in this case a program which imposes a small cost on a large number of people is quite complex from their standpoint while the benefits are simple and easy to understand from the standpoint of residents in the Tennessee Valley.

The fact that most government expenditures are not attached directly to tax measures further complicates the matter. Suppose, for example, that some project is proposed which would benefit a small minority of the population to the extent of $11.00 apiece at the cost of an increase in the government's expenditure of $9.00 per capita. The necessary tax will be quite a separate matter from the benefit. If we were to retain our assumption that individuals become informed only about matters which will probably affect them by the amount of $10.00 or more, then the prospective beneficiaries of this project would learn about the benefits, but not about the costs. If we were to assume that the minority to be benefited is 1 per cent of the population, then the cost benefit ratio of this project would be 900 to 1, yet the project would probably pass through an elected legislature since it would be favored by a significant group and opposed by no one.

We can go even further; suppose that a project which will benefit a group is very simple in its direct effects, but the costs that it imposes are quite complex. If the cost of becoming informed about the benefits were, say, $2.00, while the cost of becoming informed about the indirect injuries that it would inflict were $20.00, then it might well be accepted by the voters even though the cost was greater than the benefit. If, for example, the benefit directly conferred was $5.00 per head, while the cost imposed in a devious and complex manner was $15.00, then a voter for whom the likely net effect of the measure would be a loss of $10.00 would vote for it. Paradoxically, his vote would be completely rational. The cost of becoming fully informed on the issue, which would lead to a negative vote, would be $22.00, so the rational man, in this case, would vote against his interest because he would be investing in information in a purely rational manner.

As coauthor of *The Calculus of Consent,*[7] in which log-rolling is stressed, I am naturally interested in the likely effect of these information considerations on log-rolling. In that book perfect information models are used throughout. It is assumed that each individual properly calculated the costs and benefits which he would receive from each measure. With this information, he engaged in vote trading and achieved a result which led to too much special-interest legislation. Our present model taking information cost into account points in the same direction, but the effect is much stronger. If we were to combine the two models, the net would be a still stronger bias toward special-interest legislation, particularly since the principal restraint on such legislation in the perfect information models is the cost of log-rolling. Since this is always highly complex, the individual would normally not invest enough resources in becoming informed to understand it. Thus, log-rolling could be expected to be carried to extremes because the voters did not really understand the cost part of the cost-benefit calculation.

It might be thought that these considerations invalidate the perfect information models of *The Calculus of Consent*. This is not so. Most of the models are simple, direct democratic systems, in which the tax cost of each measure is weighted against the benefit. Clearly, such situations do exist in the real world, particularly at the local level. Since it is very easy to understand the tax cost in such situations, and the benefit of each proposed measure is also easily understood by its beneficiaries, the perfect information model is not a violent distortion of the real world. Information costs come into such situations only as there are indirect secondary consequences which may not be easily understood. We could predict that the voters would behave as the perfect information models show them behaving for the simple direct effects of the benefits and the tax cost. Only if the proposal had additional effects which were hard to understand, would the information cost models be relevant. Such secondary costs and benefits probably do exist, the effect in discouraging industry of local property taxes may be taken as an example, and our present model would indicate that they would not be properly taken into account. On the local level, however, it seems unlikely that these secondary effects are terribly important. The perfect information model would thus be a reasonably close fit.

For representative democracy, the problem is more complicated. There are two sets of people who have to be considered, the voters and their representatives in the decision-making bodies. The representatives are normally much better informed than are the voters, in fact better informed than the voters could ever be expected to be. Moreover, in the case of the representatives, the cost of becoming better informed on one subject is normally a sacrifice of information on another. If we consider a congressman as a typical representative, he devotes full time to his job, and normally the limitation on the time he devotes to becoming informed about the substance of the legislation upon which he votes, is the time he devotes to becoming informed on the wishes of the voters. We thus have a rather well-informed agent trying to carry out the wishes of ill-informed principals.

Without going into this problem too deeply, there is a simple model which fits the situation fairly well. The voter will be well informed and deeply interested in any special benefits he receives or thinks he might receive from the government. He will also be well informed on any special injury, such as an "indirect tax"[8] which falls on him. In addition, he is aware of any general tax, such as income tax, which he pays. All of these things are either so easy to learn or so important to the individual voter that the rationally ignorant person can be expected to have this information. Certain general benefits provided by the government, such as police, schools, defense, etc., are such that at least some information on them is also easily obtained. In these cases the voter is more likely to know of some scandalous deficiency than of anything else, but at any rate, some knowledge on such matters as the need for repairing the streets, school conditions, number of unsolved crimes, etc., is apt to be held by the rationally ignorant voter.

We can group these factors into two general classes about which the voter will be informed and have preferences. The general taxes and special taxes which an individual pays will both be objectionable to him. He will oppose their being raised and favor their being lowered. With respect to any special benefits he may receive, and the general benefits produced by the government he has an equally simple preference function; he wants more and he certainly does not want them reduced. To these relatively realistic assumptions about the information and desires of the voter, let us add that he knows whether the budget is balanced and that he strongly favors balance. This last would have been descriptive thirty years ago, but may not be so now.

This set of preferences and this degree of knowledge on the part of the voters would lead the representatives in Congress to engage in vigorous log-rolling with the objective of giving to a majority of the constituents in each of their districts as close a fit to their desires as is possible. The full log-rolling model, thus, can be utilized to analyze the situation. The lack of information on the part of the voters is only relevant in that they may not understand the general benefit portions of the governmental system and they may be overlooking longer range effects.[9] The voter is also basically affected by changes in the level of taxes or benefits; he does not have any clear idea of whether he is getting a good bargain in total. He cannot tell whether or not more efficient government would give him the same benefits with fewer taxes.

If the voter can hardly hope to have enough information to have any effect on the efficiency of the government, this might be assumed to merely put him in the position he is in normally in the market. He surely also does not know enough about automotive engineering to know whether the car he buys is produced as efficiently and priced as low as possible. In the market, however, there is competition, which can be relied upon to keep the producers efficient. If there were another government offering the same line of services for less, then the ignorant voter would have no difficulty in obtaining governmental efficiency.[10] The congressmen, of course, do have motives for improving efficiency since this would enable them to improve the satisfaction of their voters. Unfortunately, improvements in efficiency normally mean the discharge of government employees. Since such employees are also voters and this is a direct reduction in the benefits they are receiving from the government, there are other motives leading congressmen to oppose efficiency. What the outcome will be in any given case cannot be predicted, but the present American Congress shows little or no interest in efficiency.[11]

A basic assumption which we have been using in our discussion is that the voter objects to an unbalanced budget. Clearly, this was true not so long ago; equally clearly, it is becoming less true now. In a way, the principal effect of the "Keynesian" revolution has been to reduce the force of this objection. If this, basically irrational, preference changes, then the congressmen can solve their problems by simply spending more than the government takes in. The inflationary effect of such continual deficits is probably too complex to be understood by rationally ignorant voters. For people like myself who feel that any one of a number of methods of eliminating depressions by manipulating the quantity of money is desirable, this poses a most difficult problem. If we cannot expect the average voter to understand these systems and enforce compliance on his congressman, then the balanced budget, with all of its crudities, or continual inflation would appear to be the only alternatives.

Our model so far has been relatively crude. In order to develop a more rigorous model, let us consider a problem to which Downs devotes a great deal of attention: What is the payoff to the individual from voting? Assume that you are in possession of some information and from that information you have decided that you favor the Democratic party or, if it is a primary, some particular candidates. The payoff could be computed from the following expression:

$$BDA - C_v = P$$

B = benefit expected to be derived from success of your party or candidate

D = likelihood that your vote will make a difference

A = your estimate of the accuracy of your judgment ($-1 \angle A \angle +1$)

C_v = cost of voting

P = Payoff

Certain aspects of this expression deserve a little further discussion.[12] The B refers, of course, not to the absolute advantage of having one party or candidate in office, but the difference between that candidate and his opponent. In Downs's presentation it is called the "party differential" to indicate that it is a difference, rather than an absolute measure. A good many of the intellectuals who voted for Johnson disliked him and would have opposed him for President if he had been opposed by his running mate Hubert Humphrey. In spite of their negative evaluation of Johnson, however, they would have shown quite a high "B" value because of the strength of their detestation of Goldwater. Choosing between two evils, they nevertheless thought the choice was a most important one because they thought the two evils were of vastly different magnitude.

The factor labeled A, the estimate of the accuracy of the voter's judgment, is normally left out of discussions of this kind. It is included here because we are preparing to consider variations on the amount of information held by the individual, and the principal effect of being better informed is that your judgment is more likely to be correct. Note that it is put in the subjective form because that is the only way that a judgment can be formed, but an objective figure could be substituted without any real change in the formula. The factor labeled A can take any value from minus one which represents a certainty that the jdugments will be wrong,[13] to plus one, which indicates that the voter is sure he is right. The choice of this rather unusual way of presenting what is really a probability figure is due solely to its use in this particular equation, not to any desire to change the usual probability notational scheme. For the equation to give the right answer, it is necessary that A have a value of zero when the individual thinks that he has a fifty-fifty chance of being right.

The factor labled D is the likelihood that an individual's vote will make a difference in the election; that is, the probability that the result if he were to vote would be different than it would be if he were not to vote. For an American presidential election, this is less than one in ten million. C_v is the cost, in money and convenience of voting. For some people, of course, it may be negative. They may get pleasure, or at least the negative benefit of relief of social pressure from voting. If we view voting as an instrumental act, however, something we do, not because it gives us pleasure directly, but because we expect it to lead to some desirable goal, then our decision to vote or not will depend upon weighing the costs and benefits. For most people, the cost of voting is probably somewhere between $1.00 and $5.00.

Let us put a few figures into our expression. Suppose I feel that the election of the "right" candidate as President is worth $10,000 to me. I think I am apt to be right three times out of four, so the value of A will be .5, D will be figured as .000,000,1. Assuming that my cost of voting is $1.00, the expression gives ($10,000 × .5 × .000,000,1) − $1.00 = −.9995. It follows from this that I should not trouble to vote. This result has shocked a lot of people since Dr. Downs first presented it.[14] Since our concern is not with whether you should vote, but with information, we will not devote much time to the issue. A number of arguments have been advanced by various scholars as to why an individual should vote, but we need not go into them.

It will, however, be worthwhile to consider a few variations on the expression. In the first place, it is frequently argued that this line of reasoning would lead to no one voting. This is not true. If people began making these computations and then refraining from voting, this would raise the value of D, since the fewer the voters, the more likely that any given vote will affect the outcome. As more and more people stopped voting, D would continue to rise until the left-hand side of the expression equaled the right. At this equilibrium there would be no reason for nonvoters to begin to vote or for voters to stop. Presumably the people voting would be those among the population who were most interested in politics, since D would have the same value for everyone but (B × A) would approximate a positive function of political interest.

The equation, if it is thought to be in any way descriptive of the real world, would imply that people would be more likely to vote in close elections, and that they would be more likely to vote in local elections than in national ones since D would be larger in those cases. The first hypothesis was tested by Riker and Ordeshook[15] and found to be correct. The second would appear to be easily falsifiable. The problem is that A and B in the equation vary greatly between national and local elections. In local elections the party differential for the voter is apt to be small and the difficulty of getting information may be great, with the result that A approaches zero. As a result, a decline in voting would be expected as the political unit gets smaller, until it becomes so small that each individual knows the candidates. Here A would be large, and the probability of voting correspondingly increased.

It should also be noted that the equation is not strictly suitable for members of pressure groups. For them, the number of votes cast may well be important even if it does not change the result of the election. Insofar as the politicians are able to figure out how many votes were cast on the "farm issue," for example, they will take this bloc into account in their planning for the next election. Thus, in a sense, D is always unity. In another sense, however, D is just as small for the pressure group voter as for the man trying to elect a President on broader criteria. In each case, your own vote is only a tiny fraction of the whole. The major difference here is that the pressure group need not be anywhere near a majority of the population to be effective. Thus, instead of contributing one vote to a national majority of something in excess of 35,000,000, the member of a pressure group may be one vote in a voting block of 1,000,000. Obviously, his influence is greater in the latter case.

Our voting experession requires another elaboration to bring out its full meaning. An individual, in deciding how to vote, may take into account both direct benefits he receives and a sort of charitable benefit he receives from helping others. Thus, the B of the expression could be replaced by $(B_p + B_c)$, with B_p representing the direct benefits that the voter expects to receive and B_c representing the benefit he will receive because he gains some satisfaction from other people being benefited. To use an example that one of my students used as part of a vigorous attack upon my general position, suppose an Iowa farmer estimated the value of retaining his farm subsidies if Johnson won, at $5000, while he felt that the benefit to the United States, as a whole, from a Goldwater victory would be $10,000,000,000. (The fact that the farmer would receive his aliquot share of the $10,000,000,000 may be ignored.) If the farmer had the normal impulses for charitable activity, he would put at least some value on the benefit to the rest of the nation and, given the immense value of the benefit in this case, the value he put on it would certainly far outweigh the $5000.

This logically raises two questions: whether the farmer would put enough value on the charitable benefit to get a positive payoff from voting, and whether he would put enough value on it so that he would vote against his direct interest in the farm subsidy. Starting with the first, we may use the values used above, to figure out the value that the farmer would have had to put on the $10,000,000,000 benefit to the nation to be willing to vote for Goldwater. The answer turns out to be $20,000,000. This is the minimum size of the benefit the farmer would have had to estimate that he would receive if his investment of $1.00 in casting a ballot, which has a one-in-ten million chance of being decisive, was to have a positive payoff. This is equivalent to saying that he would have had to be willing to pay $20,000,000 if such a payment would insure a Goldwater victory. Obviously, the farmer would not have been willing to make such a payment; in fact, he didn't have the money. But the St. Petersburg paradox is involved here, as is the fact that individuals do not necessarily act in perfect accord with the computed odds. I imagine that most people, if asked whether they would be willing (if they had the money) to pay $20,000,000 to confer a benefit of $10,000,000,000 on the nation as a whole, would reply "yes." I also suspect that few of them would, in fact, be willing to do so if their fortune were not very much larger than $20,000,000.

If the farmer had decided to vote for some such reason as duty or habit, however, he might still have put the public good above his private interest without giving it an evaluation of $20,-000,000. Indeed, if he felt that the benefit he would receive from feeling that the rest of the country is well off exceeded $5000, clearly he should have voted for Goldwater if he voted. In practice, of course, he was unlikely to face such a clear problem. Decisions as to which party will do the most for the "public good" are difficult to make. Both parties will be claiming that they will make more of a contribution than the opposition. The private benefits, on the other hand, are relatively easy to work out. The farmer is likely to have figured out which side would have given him the most personally and then permitted himself to be convinced that this side was also the best for the country.

We can, however, conceive of an experiment to determine how much voters weigh benefits to other persons. There is currently much concern in the rest of the United States with the politics of Mississippi. Suppose the state of Mississippi agreed to sell a vote to anyone who was not a resident of the state, not engaged in business in the state or selling things to the state government, and was not going to move to Mississippi for five years.[16] It would thus be possible for the people who are concerned with Mississippi's state of affairs to change it by obtaining control of the state government. With the restrictions we have imposed, the only motive they would have would be charitable—the desire to benefit other people. Thus, by changing the price of the votes, we could get a measure of the strength of the desires for the "public good" of a number of people. I would guess that a lot of votes would be sold at $10.00, rather few at $100.00, and almost none at $1000.

We can vary our Gendanken experiment to produce somewhat different "results." Suppose that the state of Mississippi, instead of selling votes itself, permitted its *voting* citizens to sell their votes to persons from outside the state who had the qualifications listed above. This would not give a simple measure of the relative value that voters put upon their selfish and altruistic values, but it would permit comparison of the value that some voters (the Mississippians) put upon political decisions which are close to them and directly affect them and the valuation put upon more distant considerations by other voters. To make a guess, I would imagine that the result of this experiment would be an equilibrium price between $25.00 and $50.00, but with very few votes sold. If my guess is right, most Mississippians would feel their interest in retaining the status quo to be more than $50.00, while most outsiders would feel that their interest in changing it would be worth less. But this represents an excursion into possible methods of measuring intensity of feelings.

Returning to our problem of the effect of ignorance, let us once again complicate our model. An additional factor, C_i, the cost of obtaining information, has been included in equation (1).

$$BDA - C_v - C_i = P$$

This is, of course, the cost of obtaining additional information, since the voter will have at least some information on the issues as a result of his contact with the mass media. Of course, A is a function of information ($A = f(I)$,[17] and, hence, each increase in information held will increase A and, thus, raise both the benefits and the costs. The problem for the rational individual contemplating whether or not he should vote, would be whether there are any values of C_i which would lead to a positive value payoff.

Suppose, for example, that the investment of $100.00 (mostly in the form of leisure foregone) in obtaining more information would raise the value of A from .5 to .8. Using the same amounts for the other values as we used previously, $P = -100.992$. Clearly, this is even worse than the original outcome. Furthermore, these figures are realistic. The cost of obtaining enough information to significantly improve your vote is apt to very much outweigh the effect of the improvement. This is particularly true for the average voter who does not have much experience or skill in research and who would put a particularly high negative evaluation on the time spent in this way.

A further implication of our reasoning must be pointed out.[18] There may be social pressures that make it wise for the individual to make the rather small investment necessary for voting. In terms of our equation, C_v may be negative. In these cases, voting would always be rational. Becoming adequately informed, however, is much more expensive. Further, it is not as easy for your neighbors (or your conscience) to see whether you have or have not put enough thought into your choice. Thus, it would almost never be rational to engage in much study in order to cast a "well-informed" vote. For certain people, and presumably most readers of this book will fall within this category, A may already be quite high. For intellectuals interested in politics, the amount of information acquired about the different issues for reasons having nothing to do with voting may be quite great. Further, for this group of people, the value put upon the well being of others *may* be higher than in the rest of the population. It may be, then, that these people would get a positive payoff from voting even though the average citizen would get negative returns from taking the same action. Thus, for many of the readers of this book, voting may be rational. I have my doubts, however. The value put upon the well being of others must be extremely great. Further, my own observation of intellectuals interested in politics would not confirm that A is high for them. They may have a great deal of information, but this seems to have been collected to confirm their basic position, not to change it.

CHAPTER VIII

The Politics of Persuasion

In the last chapter we made a type of cost-benefit analysis of the individual decision to acquire information. This analysis was confined to information acquired for essentially political reasons. Information acquired because a person enjoys acquiring information, or as a sort of by-product of other activities was ignored. People do, however, have curiosity, and enjoy, to some extent, acquiring information. It is this fact which largely accounts for the existence of the media discussed in Chapter VI. It is now possible to turn to discussion of the information individuals have even if they do not engage in any special effort to be well informed on political matters.

For the purposes of our model we shall drastically simplify the variations in the amount of information that individuals may have. Instead of a continuous variation, we shall assume that for each possible political issue, the voter is either: ignorant, casually informed, or well informed. In order to distinguish between these three states of knowledge it is necessary to devote a little attention to the mass media through which most voters obtain what information they have on politics. Most of the mass media which carry any political information at all, combine it with a great deal of other material. The newspapers, for example, carry some stories of political relevance, but they normally devote less space to them than to sports and sensational crime stories. There will also be a good deal of space devoted to society news and local and national happenings, which are of no particular importance to the voter qua voter.

It must be assumed that the people who operate these mass media have a good idea of their customers' tastes. Furthermore, it seems likely that most consumers of mass media do not simply go through from one end to the other. The typical newspaper reader[1] reads only the sections that interest him, possibly scanning the rest. For a great many readers, then, only occasional political pieces will be read. Moreover, the voter does not necessarily remember for any period of time what he does read. Last, but not least, it is unlikely that he gives much, if any, thought to most of the political information he does remember.

As a first distinction, then, the voter may, at the time he votes, be in complete ignorance on many issues. This is either because he actually never heard of them or because, although he did come across them in the course of his contact with some exemplar of the mass media, he was not impressed enough to remember them. In contrast, he may be aware of the issue and have some amount of factual information about it as the result of essentially casual receipt of information together with an evaluation of that information which made him remember it. We shall call this state "casually informed," and it will be our principal concern for the next few pages. Remember that it involves both the receipt of information and the intellectual effort involved in remembering it.

Remembering the models of the last chapter, it seems likely that for most voters, knowledge on most issues never goes beyond the casually informed stage. Thus, we must assume that most voters have not thought very much about most political issues when they enter the voting booth. Also they normally will have only a little information on most of these issues. This casual receipt of information, however, does give the voter some preliminary information on quite a number of issues. If there is some reason for him to become better informed on one of them, he can do so. Again, we shall simplify and draw an arbitrary line. If the voter is sufficiently interested in some particular issue so that he devotes as much thought to it as he would to the purchase of a new car, we will say that he is well informed on the issue—we would not expect to find many such voters. Note that our standard of "information" is actually measured by the amount of thought which the voter has devoted to the subject. Since it is unlikely that he would think seriously about the subject without at least some positive efforts to increase his information, however, this is not a serious objection. Normally, people with a significant interest in a given subject do at least some investigating in order to increase the amount of information they hold on that subject.

We have not assumed that the well-informed voter is capable of making accurate judgments in the field in which he is well informed, only that he has thought seriously about it. This may mean very little for many voters. People buy things in the market which are quite badly suited to their needs,[2] and it must be assumed that they are at least as inefficient in making political decisions. Still, well-informed voters at least know more than the casually informed or the ignorant. Note, however, that our definition implies only that the voter is well informed on some single issue. A voter might, of course, be well informed on several issues or even on all significant issues, but this is not required. The voters are either ignorant, casually informed, or well informed on each issue, and this state of information is what affects their voting. The well-informed voter on one issue may well be ignorant on others.

We might now inquire why people are at some given stage of information on various issues. Everyone starts out ignorant on everything, and given the variety of subjects which we could study and the amount of information available, everyone remains ignorant on very many subjects throughout his life. The average voter, however, is exposed through the mass media to a good deal of information on political matters. Some of this information will catch his eye so that he notices it and appear important enough to him so that he remembers. In part this is obviously a simple random process; in part, however, it is more. The casual reader is more likely to have his eye caught by an item which is nearer to his interests than other items. An Irish citizen of Boston who is employed in a bank and engages in a good deal of hunting on weekends is more likely to notice and remember items on these general subjects than on others. Thus, he may normally know the names of all Irishmen who are candidates for office, will be likely to have heard of any major expenditure of federal funds in the Boston area, be interested in any possible changes in banking regulations, and keep an eye on changes in the hunting and conservation laws.

It may be fairly assumed, then, that our Irish voter will notice articles on these subjects which appear in the newspaper, and remember, in general, what they say. He will also, no doubt, notice and remember a random sample of articles on other subjects. Will he, however, feel moved to improve his knowledge on any of these subjects so that he becomes well informed? Looked at in strictly rational terms, this is unlikely. As we have shown, the likely payoff which the voter will obtain through casting a more informed vote is very small, normally much smaller than the cost of becoming well informed. In most cases a voter will be well informed only if he has some reason for gathering the information, other than the possible effect on his vote. Our voter might, for example, be well informed on the banking laws and regulations as a necessary part of his business and would naturally apply this knowledge in his political activities. At the very least, he would know enough about banking and the sources of information on banking laws so that becoming well informed on this matter would be very easy.

Similarly, the voter might live in a society where Irish politicians were very prominent both socially and as subjects of conversation, in which case he would automatically be well informed on this subject. He might also consider politics a sort of hobby, and devote a good deal of time to its study for the pleasure he gets out of it, or he may feel that it is the duty of a citizen to be well informed politically, and, thus, studies politics for a feeling of ethical justification. These last two motives, although perhaps widely applicable, apparently do not lead to much effect on the balloting. Most people who are well informed for these motives are quite firmly committed to one political position and do not shift.[3]

It might be argued that the members of pressure groups would find it rational to be particularly well informed. The argument would turn on the fact that pressure groups are normally much smaller than the total of voters, hence that each vote would be a larger share of the amount needed to get the boodle. Further, it is generally true that the absolute number of votes that can be mustered is important for a pressure group. In fact these matters are of little importance in deciding whether or not it is rational for the voter to seek out more information. Usually the potential member of a pressure group does not need much information to discover which side his bread is buttered on. The knowledge that Johnson was firmly committed to keeping farm prices up and that Goldwater proposed to adopt a gradual program of removing price supports was all that was needed for the average Iowa farmer in 1964.

It may, indeed, be better for the pressure group voter not to know much about the particular issue. Surely, the more the Iowa farmer had learned about the farm program, the more likely he would have been to have felt that there was a potential conflict between his ideas of right and wrong and his material interests. By not thinking much about the issue, he was able to follow his self-interest with a clear conscience. Similarly, the goldmining interests who were so shocked by the prominent role of Milton

Friedman in the Goldwater camp could have a simple clear identity between their devotion to their material interests and their devotion to the gold standard only so long as they did not learn enough economics to understand Friedman's criticism of that system. Even with a good knowledge of economics they might have remained believers in the gold standard, but they would have at least realized that their opponents had a case.

Further, the man who is well informed, in our rather restricted definition, on any given subject may, in fact, be very badly qualified to judge it. Bankers, for example, would be well informed about banking matters because they have given them serious thought. As every economist knows, however, they are almost impossibly bad judges of economic policy in this area. This is simply because the very large amount of information they have on the operation of their businesses normally leads them to make invalid inferences about the system as a whole. It is unfortunately true that many simple and correct principles for the management of individual banks are almost directly opposite to the correct principles for the management of the banking system. The well-informed banker knows the basic rules for operating his own business and, not unreasonably, assumes that the same rules apply to the system as a whole. This leads him into extremely foolish policy positions.

The voters who are well informed will frequently have a great deal of information which is not relevant to the specific political choice, but not necessarily much relevant information. This is because they have accumulated this information for reasons having nothing to do with the choice itself. They have acquired it in the course of their business, as a hobby, or because they enjoy politics. None of these reasons for "learning" would lead automatically to the particular set of information which would be of the most use in deciding on some particular political issue. Thus, the votes of the well-informed may not be much more accurately calculated than those of the casually informed.

The model now gives us a simple idea of the information flows. (This model is essentially the same as that in Chapter VI except that it looks at the phenomena from the standpoint of the consumer instead of the supplier.) The mass media carry various items, mainly because they think that they will sell. In this connection it should be remembered that some "mass media" are actually rather specialized in their appeal. *Vogue* will do for an example. The individual voter is (at least to some extent) exposed to these mass media and, therefore, picks up at least some information on some subjects of political interest. To this point we can regard him as being ignorant on a number of political subjects, and being casually informed on others. Presumably there is a random component in the subjects which he remembers, but, in part, this also represents the result of some rational selection. He is more likely to look at and then remember items which appear close to his interests. Among the subjects upon which he thus becomes casually informed, there may be one or more which appear to him as worth further study. Upon these he will become "well informed."

The voter who is well informed on one or more subjects is likely to cast his vote very largely in terms of these particular issues. The reason he has become well informed is his greater interest in these fields, and this greater interest is likely to carry over into the voting booth. If it does not, he will differ little in his effect upon the election from a casually informed voter. Since the only areas where an individual will be inspired by rational considerations to become well informed are the areas where he has some special interest, the well-informed voter is the pressure group voter par excellence. Looked at from the standpoint of the congressional representative seeking reelection, the well-informed voter is probably a rare phenomenon, but he is at least easily predictable. There may be a few well-informed voters, but it is fairly simple to predict how the vote of any given one of them will be affected by a given piece of legislation.

The more normal voter, who has only casual information, is less predictable. Presumably he is casting his vote largely because he thinks it is his duty, because he gets enjoyment out of it, because it is something he was taught to do, or because there is considerable social pressure to do so. In any event, he surely does not think it very likely that his vote will make much difference, and hence does not regard his decision as being worth a lot of thought and study. Under the circumstances, he presumably inspects the information he has in his memory, and casts his vote in view of these various "facts." Let us, for the time being, assume that on each issue upon which the voter has any information, he knows, among other things, the stand taken by each candidate. Thus, he simply adds up the stands taken by the candidates on various issues in terms of his own opinion and the weights he gives to each policy problem, and decides how to vote. This means that the politician must take into account, for each issue, the likelihood that voters will be even casually informed about it, the policies which the voters will favor, and the relative weight they will put upon it in making up their mind in the polling booth.

In part, the information held by any voter is random and the politician can only assume that this information is evenly distributed through the population. (Some information held by the individual, because it affects his interests, may appear random to the politician, and he is unlikely to know much about who has what hobbies.) In part, however, the casual information of any individual reflects his relative concern with different problems, and here the politician will have a somewhat better idea of what he will know. The Irish will know if Irishmen have recently been put in high office and the bankers will know about new banking legislation, even if neither has more than a small amount of information on these points. It is fairly easy to guess what they will think about certain types of problems too. Those inhabitants of Richmond who know about the project to dredge the James River probably all favor it. This may, of course, be a mistake on their part. Surely some citizens of Richmond will gain markedly from this project, but it is by no means obvious that the gain made by an average citizen will exceed the tax cost to him, even if the tax is assumed to be distributed over the entire country. The individual benefits for most of the residents of Richmond, who are not directly connected with the project, will be small and indirect. Being casually informed about the existence of the project, but never having given the matter any serious thought, the individual is likely to have a vague idea of the benefits to be expected, but no idea of the cost. Thus, he may count it as desirable even if it would, in net, injure him.

The voter, in weighing the various issues, is likely to give the greatest weight to those closest to him, just as he is most likely to know about them. Thus, it seems likely, again, that the casual voter will act like a "pressure group" voter. He will differ from the classical idea of a pressure group voter only in that he may either be ignorant on any given subject or badly informed. These defects in the voter's information add a stochastic element to the equation. The congressman can never be sure when he will get an especially bad distribution of information among the voters, with all the things which will count against him known and all that would count for him unknown, but this merely makes his profession a dangerous one. In addition to increasing the risk, however, the existence of large numbers of ignorant and casually informed voters changes the parameters of the problem for the politician. He is, in the first place, less interested in the likely effects of a governmental action on the voter than in the view taken of the matter by the voter. The ignorant voter will take no view at all, so he can be ignored. The casual voter may be completely incorrect in his assessment of the problem, so the vote maximizing action may be directly contrary to his interest.

The pattern of behavior which this picture of the information held by the voters dictates for the politician is essentially that described by log-rolling. If there are any political problems on which there are widely and strongly held opinions, then he should try to follow those opinions. In addition, he should try to find opportunities to do things which will confer a simple, easily perceived benefit on small groups, but whose cost is dispersed and hard to understand. But he may be able to do better by manipulating the information held by the voters. Since this information is casually acquired, it may be possible to make it likely for the voter to casually acquire information favorable to the politician. The political campaign, the making of speeches and appearances, all have this effect. The dependence upon the real lack of voter interest in the issues, which characterizes most politicians, can, perhaps, best be illustrated by the fact that many politicians will not reply to charges made against them because they feel that this would give the charges a wider currency than they would otherwise receive.

In addition to these methods, advertising in the more conventional sense may be a big help to the politician, and most politicians engage in a good deal of it. They may also be able to influence the content of the mass media. In general, the editors of these sources of news are interested in obtaining the largest possible revenue with the smallest possible outlay. Under the circumstances, they are apt to be aiming at reducing the cost of their editorial matter, and the politician may be able to take advantage of this by providing suitable editorial matter at little or no cost. Furthermore, by regularly providing favors to the mass media, he may get them to give his "side" prominence and suppress information which might cause him political injury.

It must not be forgotten, however, that the politician is not the only one trying to influence the information held by the average voter. He will have an opponent who is trying to counter his strategies. This limitation means that the politician is unwise if he does anything which will seriously injure any important group of voters. If the injury is great enough so that it would pay his opponent to bring it to the voters' attention, he probably will. Thus, again, the politician should aim at policies which confer easily seen benefits and dispersed and hard to understand costs.[4]

So far we have been assuming that the voters know what the stands of the various candidates are on the issues. There is no particular reason why the casually informed voter should have any better information on this matter than on any other. The old lady in Vermont, who, because she was convinced that Barry Goldwater was opposed to TV, changed a lifelong habit of voting Republican, is merely an extreme example of the ignorance we may find in this area. This is, of course, one of the reasons for backing a party instead of individual candidates. It is easier to know what a party stands for than to know what each of a long list of candidates stands for. The casually informed voter may quite frequently be misinformed on this point, but he no doubt does have a better idea of what the party position on a given matter is than of what the individual candidates advocate.

In primaries, the voter does not have this crutch. He must make up his mind among candidates who receive little attention in the mass media, and who all appear to accept the positions of a given party. Under the circumstances, it is not very likely that he will have very good information on the stand of the candidates on those issues upon which he has made up his mind. Surely there must be an extremely large amount of ill-informed voting here. In fact, the democratic process may well have a basically stochastic effect at this level. It is also an area where public relations skill is of overwhelming importance.

Still, the expected pattern of behavior we would deduce for a politician would involve a great dependence upon log-rolling. The principal change which our limited information models make in the full information models of *The Calculus of Consent* and *Entrepreneurial Politics*[5] is to introduce an element of improper calculation. Further, since the limitations on information do not have an entirely symmetrical effect, the misallocation of resources to be expected under log-rolling is magnified. In justice, however, it should be pointed out that the principal advantage of log-rolling, the protection of minorities against exploitation, is also magnified by the ignorance present in these models.

On the basis of this relatively realistic view of the political information held by the individual voter, we can now discuss some problems of political persuasion. This discussion will, naturally, have some similarity to the media discussion in Chapter VI, but basically it will be concerned with individuals and will use variants of the equations developed in Chapter VII.

There is, however, another way to affect our government. Instead of worrying about our own vote, we may attempt to influence others. In equation (2), C_p is the cost of effort invested in persuasion, and D_p the likely effect on the outcome of the persuasion.

$$BD_pA - C_i - C_p = P$$

It is by no means certain that efforts to persuade will have a negative payoff even if voting does. In the first place, the number of people trying to persuade is a very small fraction of the number of voters. Since the average voter doesn't do a great deal of personal thinking about politics, and picks up his ideas from the mass media, it is reasonably certain that D_p in equation (2) for the average "persuader" must be very much larger than D in equation (1). Putting the same thing another way, advocacy is more likely to affect the outcome of the election than is voting. Needless to say, this is only for the average persuader. There will be very great variation in D_p from person to person. In some cases it may even be negative.

Another difference between equation (2) and equation (1) is that C_p is very small for some people. Persons engaged professionally in providing material for the mass media may be able to put a considerable "persuasive" effect into it with almost no cost. In some cases, indeed, it may have a negative cost. Consider a newsbroadcaster who must fill a half hour and who feels strongly that it is his duty to show the "truth." If "truth" for him means the position taken by one side, then he will obtain positive pleasure out of inserting it in his newscast. Similarly, the author of a TV series script may get a feeling of moral justification out of selecting his topics and treatment so that they "improve" the political ideas of the viewers.

The management of the mass media will put some limits on this process, of course, since it might reduce sales. In general, however, they do not care very much about the political position of their output.[6] As long as the writers do not offend too many potential customers, they will be left alone. In addition, the managers may feel the controversy is likely to sell, and actively encourage the taking of strong stands without much concern for what these stands are. Given that the average man gets almost all of his political information from the mass media, it is clearly true that individuals employed to produce material can have a very great influence with very little cost. Thus, for this small portion of the public, at any rate, the P in equation (2) would normally be positive.

This phenomenon is not limited to the producers of material for the mass media, however. At the lowest level, strong support of some political activity of the sort that involves membership in committees, local party leadership, or the signing of political manifestos is normally undertaken largely as a hobby. The individual gets a positive enjoyment both from the activity itself and from a feeling that he is "doing his duty." In any event, such activity is not irrational. It should be noted that it is not necessary to devote much in the way of resources to obtaining information. Most political activists, in fact, have more information than the average man, but not necessarily very much more.[7]

There are a great many people who are able to work political advocacy into their normal work without any cost or with small cost. In addition to the mass media, we have "class" media. The writer for the *New Yorker* or the *New York Review of Books* can put a good deal of political content into his output without annoying his editors. They may, indeed, require that he do so. Some journals, such as the *New Republic* and the *National Review*, exist solely for the purpose of political advocacy. Teachers also are in a good position to push their political views with little or no cost. This is particularly true of college teachers in literature or the social sciences who can hope not only to influence their students but also to influence other people through their students. The political science major is likely to be an opinion former in later life and thus spread his professor's ideas to a number of other voters.

Whether this sort of advocacy will, or will not, have costs will depend to a considerable extent upon both the status of the teacher and his positions. Advocating the position which is held by the overwhelming majority of his departmental seniors and members of the profession at other schools is likely to help rather than hinder the advance of an assistant professor. If he happens to hold views which are regarded simply as wrong by his immediate superiors, he is unlikely to be retained. Thus, the cost to an individual who happened to favor Senator McCarthy and was an instructor in an Ivy League School in the early 1950's would have been very heavy.[8] Even with tenure positions, a sufficiently great deviation can be costly. But, in any case, giving his opinion to his students will not involve any extra work to the teacher and is thus, in this sense, costless.

Those teachers who teach potential teachers have even greater leverage. Their ideas may, after a number of years, be spread to thousands of voters. The time lag, however, must be kept in mind. It is probable that only the people who work in the mass media can have any significant effect on an upcoming election. The writers for the intellectual journals and the university professors will have sizable effects only after a considerable delay. Since there will necessarily be this delay, it may be that they will be less interested in strict partisan politics than in trying to spread certain ideas or philosophies. As has been discussed in earlier chapters, the political parties adjust themselves to the spectrum of preferences in the society, and changing this spectrum may be a more effective way of influencing developments than developing party loyalties. The situation can be shown by equation (3):

$$A \times S - C_i - C_p = P$$

"S" is the amount of shift in "opinion" on some given issue, which is the result of the persuasion measured in terms of the "utils" obtained by the potential persuader.

Clearly, for people who have relatively low C_p's, this type of action can have a high payoff. Speaking personally, I am interested in improving the rationality of the political system under which I live, which is a complex of ideas. By writing on the subject and by teaching my views, I clearly have at least some effect in that direction.[9] Since I am paid to teach and write, and there is really little restriction on my choice of subject, the cost of my activities is substantially zero. Under the circumstances, P is almost certain to be positive if I give any sizable value to S and do not evaluate A at a low figure.

Diagrammatically, we can show the situation by Figure LXVI. On some issue the population is distributed as shown. The two parties, near the center of the distribution, are at R and D. The persuader changes the distribution of the population to that shown by the dotted line. The two parties, conforming to the now existing distribution of public opinion, move to R′ and D′. Since one or the other party will be in power, the government's policy has been moved to the left. The persuader can feel that he has had a

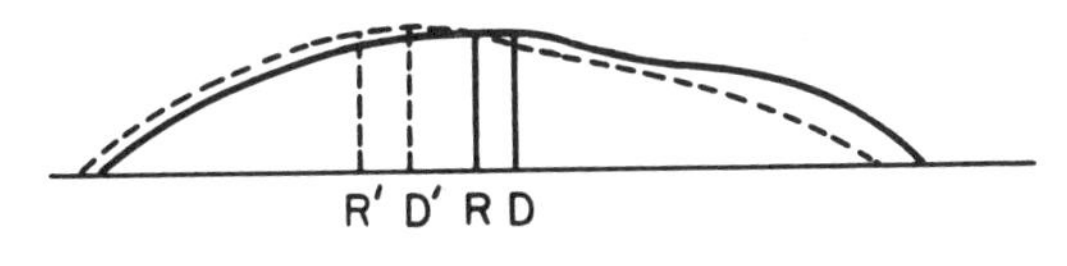

FIGURE LXVI

very real effect on "politics" even though his work may have had no effect upon which party is in power.

Needless to say, the mass media also can devote their attention to shifting public opinion rather than helping one party or another win. The television story writer, for example, who shows almost all businessmen as grasping and self-centered, while almost all of his politicians and civil servants are self-sacrificing and only interested in the public good, will surely have an effect on the "climate of opinion." Similarly, the writer of light fiction for the women's magazines who occasionally introduces farmers who are vacationing in Florida on the money they receive from the taxpayer for not farming, will probably have an effect on the public attitude toward the farm program. In a sense, the whole purpose of "public relations" is to exert this sort of influence through the mass media.

So far, in discussing the problem of persuasion, I have more or less ignored the costs of obtaining additional information for this purpose. It had been rational to engage in persuasion because the costs of doing so are zero or near zero to the individuals I have discussed. The costs of improving one's information in order to increase the accuracy of the decision as to what to advocate is not likely to be either zero or near zero. Thus, it is perfectly possible for an individual to be in a position where it is rational for him to try to persuade others of the correctness of his views, but not to engage in any research to find out whether those views are accurate. General experience would indicate that many people have drawn this conclusion. A great many people engage in rather vigorous advocacy of their positions without much effort to improve their information on them. Furthermore, as a bit of casual empiricism, most of the "information gathering" done by people who are engaged in persuasion seems to be aimed at improving their efficiency as persuaders, not at obtaining information which might lead them to advocate some other position. Such magazines as *The Nation* and *The National Review* serve the function, mainly, of a sort of "Agitator's Notebook" for their faithful readers. The liberals who read *The National Review* or the conservatives who read *The Nation*, are few. Neither group really wants information which might lead it to change its mind.

This relative unconcern for information by "persuaders" theoretically should be pretty general, and, to my mind, the theory fits the real world. In writing to persuade, inaccuracy in your information is only important if the person whom you are trying to persuade knows or is likely to find out that you are in error. Thus, you need only avoid factual misstatements which will be detected by your relatively ill-informed audience or which will be of use to the people trying to persuade in the other direction. There may not be anyone on the other side, and public ignorance may be deep and pervasive. A dictatorship will normally make public ignorance one of its main objectives. Even in a free society it may be possible to drown opposition in a sea of words. Milton Friedman has pointed out that 98 percent of all published work on the Federal Reserve System emanates from the Federal Reserve System itself. Under the circumstances, it is not surprising that its public image is a strong one.

So far we have been discussing the costs and benefits for anyone interested in attempting to change the world by political methods. There is, however, a strategic decision which also must be made. An individual can vote, try to influence voters directly, try to influence voters indirectly by influencing people who will influence other people, or he can (directly or indirectly) try to change the climate of opinion with the objective of shifting the point upon the spectrum at which the parties will locate. Depending upon his situation, any, all, or none of these actions may be rational.

In order to consider the advantages and disadvantages of direct or indirect persuasion, let us consider an abstract model of the transmission of ideas through society. In Figure LXVII we have a society of ten individuals each of whom holds a separate idea in Period I. The ideas are denominated by the first ten letters of the alphabet. Each of the individuals tries to convince the others of the truth of his idea and the result in Period II shows ideas A, G, and I, have been eliminated.[10] In Period III, B and H drop out and in Period IV ideas C and E assume dominant positions. It would be fairly easy in Period IV to predict that the eventual outcome would be the adoption of either C or E by the majority of the population. In Period V, E has obtained a majority and if it involves a political policy, it is reasonable to assume that it will be adopted.[11]

	1	2	3	4	5	6	7	8	9	10
Period I	A	B	C	D	E	F	G	H	I	J
Period II	B	C	C	D	E	F	F	H	J	J
Period III	C	C	D	E	E	E	F	F	J	J
Period IV	C	C	C	E	E	E	E	F	F	J
Period V	C	E	E	E	E	E	E	F	J	J

FIGURE LXVII

What has all of this to do with the choice between direct and indirect persuasion? At any given time there are groups of ideas which are similar to the group held in Period IV. That is, most members of society hold one of a few positions and it is fairly certain that one of these few positions will be adopted shortly. The person interested in affecting the immediate outcome is more or less compelled to adopt one of the "popular" positions and argue directly to the voters for it. There will be other groups of ideas, however, which are each held by relatively few persons.[12] At any time, then, there are ideas which can be said to be in any one of the periods of Figure LXVII. For those ideas which are currently dominant, or which appear to be very close to achieving dominance, direct persuasion in the mass media is the most suitable procedure. For the ideas which are held by few persons, however, an indirect strategy is necessary. These ideas can become dominant only if people now plan to persuade others who will persuade others, etc. If you choose an unpopular idea you must plan not for immediate success, but for long range influence.

There are several reasons why you might choose an unpopular idea to back rather than one which is now held by enough people so that it seems likely that it could be converted into a dominant idea in the next period. In the first place, since there are a great many unpopular ideas and only a few, ex-definition, which are held by large numbers of people, it is more likely that the idea which most appeals to you will be unpopular rather than popular. (This assumes, of course, that your preferences are not closely correlated with those of the average man.) Thus, unless you happen, by coincidence, to find that your first preference idea is one which is on the verge of becoming dominant, you will be confronted with a choice between an idea which you much prefer but which is unlikely to be widely adopted for some time and an idea which you like less but which may be adopted very soon. As an example, consider person 10 in Period IV of Figure LXVII. He finds that J is the idea which he likes best, but he feels that J will surely not be adopted until Period VIII at the soonest. It is obvious, however, that either C or E will be dominant in Period V; and we can assume he prefers C of those two. Thus, his decision as to whether to push C or J (he may be able to push both) is largely a choice between working for his maximum preference, obtainable some time in the distant future, or seeking to make a less significant improvement in the immediate future.

Obviously there can be no general solution to this problem. It depends, in part, upon the strength of the individual's preferences and, in part, on his estimate of the likely future developments. We can, however, consider the factors which must be taken into account. It is sometimes said that only the decision to push an idea such as C is "practical." Partly this view is based on simple blindness about possible long-run effects, but partly it reflects the fact that the odds are against any given idea now held by a tiny minority.[13] There are a great many such ideas and the odds against any one of them achieving dominance are correspondingly large. Thus, in Period I the odds against E achieving dominance at any time were about ten to one, while in Period IV the odds against C were not much more than fifty-fifty.

There are two possible answers to this argument. In the first place, persuasion in favor of some given idea may not result in that idea being adopted, even in the distant future, but it may lead to the idea ultimately adopted being closer to the one originally favored than it otherwise would be. In Figure LXVII, the people who originally argued for D did not get their desires, but it is possible that if they had not taken this position, argument F would have won out instead of E. Second, and more important, the individual should not be interested in advocating a winning idea, but in maximizing the effect of his persuasion. It is by no means obvious that the man who argues for an idea held by a large group of people and who then sees it adopted is doing more to move society toward his goals than is the man who argues for a position held only by a tiny minority and, after a similar period, sees that the minority is less tiny. In both cases, the problem is "What difference would it make had the individual not exercised his persuasive powers?" and it is more likely that the widely held idea would have been adopted than that the tiny minority would have grown.[14]

But, again, the answer to the problem depends upon the particular circumstances of the case. Liddel-Hart once said that attacks on the enemy's lines of communication had more rapid effects if they occurred close to his battle line, but that the eventual effect was greater if they occurred far to the rear. It is possible that this is also true in politics. In any event, I myself have chosen to advocate ideas which are now held by few people, in the hopes that I will be able to convince people who will then convince others, etc.

There is another strategic variable, however. It may be that political changes in normal times (emergencies may provide an exception) must always be made in a series of small steps with the voters concerned mainly with each step, not with the ultimate objective. If this is so, then it is useless to recommend a major change unless a small step in the same direction was also desirable. In order to clarify the point on Figure LXVIII, the vertical axis shows the "desirability" of various measures according to some evaluation system. Along the horizontal axis, possible states of some social variable are shown. The status quo is A, and it is assumed that society can only make changes by steps of one fourth inch as shown on either side of A. Under the circumstances, it would be possible for society to reach the suboptimum B by a series of small steps each of which was an improvement, but it could reach the optimum optimorum only by taking a big step or a series of small steps which, at first, would each make things worse. Given our assumptions about the type of actions which are possible, it would be sensible to argue for movements to the left, with the objective of making some improvement by reaching B rather than wasting time on the impossible task of arguing for movement to C.

But why should we make these assumptions about the type of political change which is possible? The answer is that if the reasons are not deducible from what has already been said about voter information, they require only a few additions. Let us suppose, reasonably, that the complexity and length of the arguments necessary to convince a man of the necessity of some social change are roughly proportional to the size of that change. From our previous work it seems likely that few voters would be willing to concentrate on an argument for some change for very long. In consequence, it would not be possible to convince them that moves to either B or C were desirable, but they might be talked into a

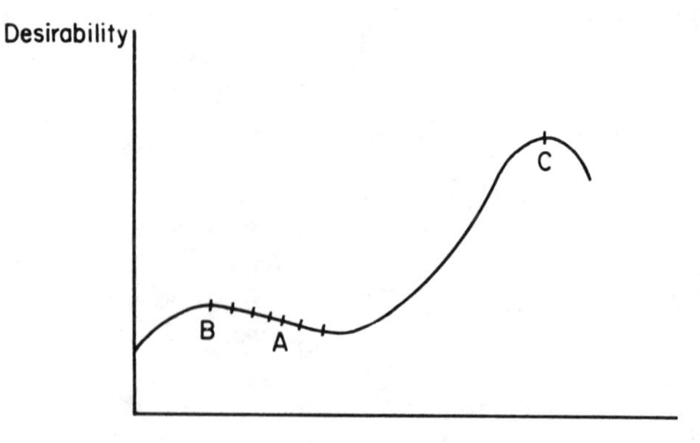

FIGURE LXVIII

single step up the hill toward B.[15] Accepting this argument, then, the types of social improvement which can be advocated with reasonable chance of success are restricted. Society could only move up to the nearest peak, and could not cross a valley in order to climb a higher mountain.

CHAPTER IX

The Economics of Lying

So far we have discussed the provision and consumption of information with only occasional attention to the fact that much of the content of political persuasion is deliberately deceptive. This chapter is intended to remedy the deficiency by discussing the use of lies and deception. It will be shorter than most of the preceding chapters, but this will not be because the subject is relatively unimportant, but because I have been unable to develop the analysis above the elementary level.

Let us begin by considering a political propagandist, a man or organization which aims to change public opinion in order to obtain some political objective. Some of the results of our analysis will be superficially paradoxical, but most of them will be in good accord with the actual behavior of the lobbyists and the public relations counsels who play such a major role in our political life. As one example, books on rhetoric normally urge placing great emphasis on the strong points of your argument while skimming over the weak. The pressure group may be well advised to do just the opposite if the weak argument will appeal to large numbers of people who are only vaguely interested in the subject, while the strong argument will appeal to a small but highly motivated group. Thus, in arguing for the farm program, its alleged benefits for the public in general, for the poor, and for national defense, etc., are much stressed. These benefits are all vague, weak, or frankly fraudulent. The benefits for the farmer, which are direct and strong, are not played up in the propaganda. The reason for this pattern is clear. The average voter does not have very much interest in becoming well informed about the program, but may well pick up an argument for it from the large volume of "general interest" propaganda. He is unlikely to be impressed favorably by the fact that the program will benefit the farmer, but if this fact is not stressed, it will probably escape his notice. The farmer, on the other hand, is motivated to find out about the program for material reasons, and is, thus, likely to look into it enough so that the material benefit for himself becomes clear.

Suppose, for example, that the average voter will, in the course of a year, see ten pages of material put out by the farm interests, and the average farmer will see about one hundred. If 98 percent of the output of the propaganda mill is devoted to "public interest" arguments for the farm program, and 2 percent to pointing out its sectional importance for the farmers, then few of the nonfarm voters will see the 2 percent, but substantially all of the farmers will. Furthermore, the farmers are likely to devote much more attention to the 2 percent dealing with direct benefits to them than to the remaining 98 percent of the output, while those nonfarmers who happen to chance upon this 2 percent in the course of their reading, will normally not give it any more attention than the other material which argues for the program on broader grounds. Thus, even the minority of the nonfarm voters who chance upon the special interest arguments will probably think of the program as mainly aimed at public rather than private goals.[1]

This principle has wide application; in fact, it permits a sort of honest deception. A politician can (and most of them do) underplay his promises to specific groups and heavily emphasize his appeals to "the public interest." It is also possible to make statements which are interpreted by each of two conflicting interest groups as supporting themselves. The dangers of this process, of course, are also obvious. In a campaign, the politician has an opponent, and the opponent has a strong incentive to bring the first politician's statements strongly to the attention of the groups they are most likely to alienate. Pressure groups are normally free of this check upon their activities. As a general rule, we do not have a set of directly opposed pressure groups. The glove manufacturers want an increase in the tariff on gloves, but there is no specific group which finds it worthwhile to organize a counterlobby.[2]

Let us, however, not confine ourselves solely to politics. Lies are found in all spheres of life. Adam Smith thought that businessmen would seldom take advantage of their customers because of the discipline of continuous dealings. In modern times, the contrary impression, that businessmen will cheat if not forcibly prevented, is, perhaps, more widely held. If we look at the real world rather than the literature, we find the widest possible variation in honesty of businessmen. The stockmarket works almost entirely by oral communication, with only the most casual written records. Yet, fraud is almost unknown. At the other extreme, certain businesses, like the provision of patent medicines, are almost purely fraudulent. The same diversity of both reputation and actual performance can be seen in politics. Politicians are simultaneously thought to be dishonest and to be people upon whose word you can rely.

Scientists are a particularly interesting group because they produce a large volume of literature of remarkable accuracy although they are not particularly noted for honesty or accuracy when they get involved in nonscientific matters. The company which used to make Carter's Little Liver Pills (now called Carter's Little Pills as a result of a rather belated awakening of the FTC to the fact that the pills have nothing to do with the liver) fell into the hands of an aggressive businessman who sharply stepped up their more or less fraudulent advertising—which led to the difficulties with the FTC. He also decided to improve his business in other ways, however, and to this end established a laboratory. Needless to say, the reputation of the company was not such as to attract leading scientists. As a result, he ended up with personnel of very little reputation. This rather ill-omened research laboratory made the principal medical advance of the 1950's—the discovery of the tranquilizers.[3] The result was that the same organization was simultaneously selling a quack nostrum by quite dishonest TV commercials and a drug that was a major medical advance, by rigorously accurate reports in the scientific journals.

Obviously, there must be some reason why people seem to be strictly honest in some situations and not so in others. It would be rational to lie if the anticipated benefits exceeded anticipated costs. The general relationship can be shown in the following fourth equation:

$$B_1 - C_1 = P_1$$

B_1 = anticipated benefits from lying
C_1 = anticipated costs of lying
P_1 = payoff

Although the equation, as it stands, will not help us much, it can be developed into a more meaningful form. Before adding to the equation, however, let us exclude certain factors. There are many reasons why one might lie, ranging from the polite lies of normal society to sheer lunacy. For the rest of this discussion, let us confine ourselves to lies told by individuals within their professions. This is not because such lies are more important or more interesting than other lies, nor because other lies would not fit our equation, but simply to make the scope of the discussion manageable. Thus, only lies told to assist in making a living will be covered by the remainder of this essay.

As a further restriction, only a deliberately untrue statement will count as a lie. This may seem to be laboring the obvious, since this is what the word means in ordinary speech, but we do not normally know for certain that when an individual makes an untrue statement, he knows it to be untrue. Further, the effect of a given statement is, presumably, not affected by the belief or lack of belief of the person who utters it.[4] There is a whole spectrum of mental states which a man making an untrue statement may have. He may be honestly mistaken; he may be mistaken, but his mistakes may always be such as to advantage himself; or he may be incapable of distinguishing between what is true and what is to his advantage. In restricting ourselves solely to the situation in which the false communication is made as the result of deliberate calculation, we are ignoring much economically relevant behavior. Once again, the only defense I can offer of this exclusion is that it keeps the scope of the investigation small enough to be manageable. Fortunately, the curious reader will find it fairly easy to extend the analysis to the cases we have excluded.

Returning to our equation, let us try to give the basic concepts more meaning. The anonymous author of *A Practical Guide for the Ambitious Politician*,[5] discussing one class of deliberate lies said "In calumny, two things are to be observed: The first is, is it sufficient. . . . The Second, is it probable?" This neatly lists the two problems involved in lying, which we may summarize as the likelihood that the lie will be believed and the probability that it will persuade the hearer to take the desired action. The lie will always be part of an effort to persuade some person or persons to take some action or to refrain from some action. It is a persuasive effort, and its benefit will come from the success of the persuasion. This being so, we can expand equation (4) to equation (5) by substituting these three factors for B_1.

$$BLP - C_1 = P_1 \quad O<L<1 \quad O<P<1$$

B = Benefit expected to be derived from the action being urged

L = Likelihood that the lie will be believed

P = Persuasive effect of lie; probability that the lie, if believed, will bring about the desired action.

Since the lie may be addressed to more than one person—in politics it may be addressed to millions—we should have a set of B's, L's, and P's for each of them. For simplicity, however, I will develop the equation solely in terms of one person although I will frequently discuss situations in which deception of more than one person is desired. Equations which dealt with many people would be much more elaborate in appearance than the ones I intend to use, but in principle they are merely summations of a set of individual equations. It seems unnecessary to confuse the issue by complicating the equations in this way.

The costs also can be presented in a more detailed form, as equation (6) below.

$$BLP - C_c - (1\text{-}L)(C_pL_p + C_rL_r) = P$$

C_c = Conscience, internal cost of lying
C_p = Costs of punishment
L_p = Likelihood of punishment if lie is not believed
C_r = Injury to reputation through other's knowledge that an individual has lied
L_r = Likelihood that injury to reputation will occur if lie is not believed

Some of these costs require a little discussion. C_c is the "pain" of doing something which you think you should not do. It, presumably, results from indoctrination in various socially approved ethical principles, primarily in childhood. Assuming, as I think we can, that the existence of lying is a factor reducing efficiency of the social apparatus, it is rational to try to strengthen this indoctrination. Thus, it would be sensible to devote resources to "moral education" for children and to try to reinforce this indoctrination among adults. On the other hand, for each individual the conscience has a negative survival value. The man for whom C_c is infinite will never be able to lie no matter how much could be gained thereby. This means that he should, in the long run, do less well in obtaining worldly goods than those for whom C_c is a distinctly minor item. Thus, if you want your child to do well in the world, you should advocate generalized moral instruction against lying while privately telling the child that lying is alright, provided only that the liar is not caught.

The conscience pain will be felt simply as a result of lying, the other two costs will be felt only if the lie is unsuccessful, i.e., if it is not believed.[6] In consequence, I have multiplied the individual costs and their individual probabilities by the probability of the lie failing. Society sometimes provides formal punishments for lies, particularly in the case of what the law terms fraud. Other examples include the honor system at West Point, impeachment, etc. If the particular lie is such that it might be covered by such a punishment system, then there will be some probability that the liar will be punished if he is caught, and C_pL_p is the product of that probability and the cost of the punishment. Since the punishment and the probability that it will be imposed are socially determined, we have here another mechanism by which lying can be reduced. If $(1\text{-}L)C_pL_p$ is made greater than BLP by either making C_p very great and/or investing large resources in "police" activities so that L approaches zero, then no rational man would lie.[7]

C_rL_r measures a more informal cost which may be inflicted on the liar whose lies become known. If it is known that he has told a lie in this case, then he may be suspected of telling lies on other occasions, and this may impose substantial costs on him. Note that this suspicion of his work might affect the particular transaction in which he is lying. If a man is trying to sell something and feels that five "factual" statements will help in making the sale but only four of these statements would be true, then he must make a rather involved calculation about whether he should or should not lie about the fifth. If he tells the lie, and it is believed, this improves his chance of making the sale. If he tells the lie and it is not believed, then the information about that particular factor held by the potential buyer is the same as it would have been if he had not lied. The buyer also now knows, however, that the seller is willing to lie if there is a prospect of profit, and hence may disbelieve his other statements, which may make the sale less likely. All of these factors affect only this particular sale. But the potential seller may have to deal with the same customer again, and he should be interested in what the customer thinks of his honesty in these future transactions. Finally, the fact that he tells lies in his sales talks may become generally known, which will also affect his future sales. Note that this particular damage to reputation is quite different from the formal punishment. Potential buyers are refraining from buying, not in order to injure the liar, but simply because they doubt whether they will get a good bargain.

Considering only the $C_rL_r(1\text{-}L)$ part of the equation will give us some information about the areas where lying is or is not likely to be a normal part of the professional activity of individuals. First, and most obviously, the more expert the customer, the less likely that a lie will be believed, hence lies will be less common when dealing with experts. The importance of the transaction will also be significant since the more important, the more likely that the potential victim of the lie will make an independent investigation and hence find out that the lie is a lie. Also, the more important the transaction, the more likely that the potential victim will inquire about the reputation of the potential liar, and hence the more important that reputation is. Most important of all, however, is Adam Smith's suggestion, repetitive dealings. It would be very stupid indeed to cheat a man once if the total profit you can make is small compared to the business you may do with him in the future.

From all of this we can deduce in a general way the likelihood of lying in various professions. Consider the salesman of some industrial product such as steel. He will be dealing with rather expert purchasers who buy in fairly large lots so that it is sensible for them to give careful consideration to each transaction. Moreover, even if they are deceived at the time of purchase, they are almost certain to find out about the deception shortly when they put the steel to use. Not only will the salesman be calling on the same customers in the future, the users of steel know each other and engage in a lot of trade gossip, so any successful cheating of one customer is almost certain to become known to others. Under the circumstances, the salesman would be most unlikely to lie to his customer. It is quite possible that the incentives for honesty may be so strong that the customer may depend upon the salesman as one of his major sources of technological information.

At the other extreme, consider a door to door salesman of some trivial gadget which does not cost enough so that the housewife will devote much thought to her decision to buy or not to buy. If the salesman does not intend to return to the same neighborhood, all of the terms in $C_rL_r(1\text{-}L)$ may have substantially zero values. Thus, any reluctance to lie on his part would have to be based on his conscience or upon the likelihood of punishment. Most commercial dealings probably lie somewhere between these extremes. The man who confines his dealings to merchants of reputation, who therefore have a reputation to lose, and deals repeatedly with the same store probably has little chance of being lied to. In "one shot" transactions or in dealing with people whose reputation is not established, on the other hand, it would be wise to take precautions. Note that this does not apply only to direct salesmanship. The considerations involved in deciding whether to run a dishonest advertisement would be much the same as those in deciding whether to lie orally. The only major difference would be the fact that an ad leaves fairly certain evidence of the lie, while an oral statement can always be denied.

Turning from general business to specialized trades, journalism is an interesting case because a false statement can readily be given much wider currency than the truth. If a newspaper chooses to lie about some subject and refuses to publish a correction, then for the average reader of that newspaper the L is high and the $(1\text{-}L)C_rL_r$ low. For some readers, specifically those libeled, L is very low and good social policy would suggest that they be permitted to make C_p high. The law of libel and slander, of course, is an expression of this line of reasoning. Interestingly, Brazil takes a completely different tack. Newspapers cannot be sued for libel, but they must print a reply from anyone they have attacked. This totally eliminates the C_p but makes the value of $(1\text{-}L)C_rL_r$ rather high. It would be an interesting research project to find out which of these techniques puts the most pressure for accuracy on the press.

Probably the most "honest" field, in terms of accuracy of publication, is science. The reason for this accuracy is fairly simple. "B" is small for minor scientific discoveries, but high for major ones. A false announcement of a cancer cure, if believed, would surely get me the Nobel prize. But this variation in B is countered by an opposite variation in L. I could probably get away with a false "discovery" on some highly obscure subject, but this would bring little reward. My false cancer cure, on the other hand, would be known to be false almost immediately. The fact that scientists repeat experiments, and naturally are most likely to repeat important ones, and the fact that new discoveries are promptly applied by other scientists, which will normally turn up any falsification, means that the likelihood of getting away with a lie of importance in this field is substantially nil. Since the benefit from a lie of no importance is also substantially nil, lying doesn't pay.[8] It should be noted, however, that there are certain areas where a scientist might find lies helpful. The senior scientist who claims credit for a discovery actually made by a junior man in his laboratory might well get away with it. In conversation with scientists you will not infrequently hear gossip about this form of dishonesty, although I cannot say how accurate the charges are.

When we have a dispute of some sort, we may refer it to an "impartial" person for decision. "Impartial" in this case normally means simply that B is zero. The person or persons chosen to decide are supposed to be so removed from the issue that he or they will derive no personal gains of any sort from their decision. The problem of conflict of interest is important only in this context. Judges, juries, and many government officials are supposed to be free of any connection with the matters they decide. With B as zero, they would incur costs but no benefits from lies. The problems which this technique raises are largely ones of motivation. The judge will normally not be directly motivated to bring in a false decision, but he may also not be motivated to give the matter enough consideration so that his decision is very accurate. The method reduces lies, but increases random error. It may also not really get rid of the motive to lie. B may be very small but not zero, but the costs may be even smaller. This is particularly true if the "impartial" person is given the contempt powers of judges to punish persons who circulate stories of his dishonesty. Still, the method has been very widely used, and there is no reason to doubt that it is valuable in many situations.

The politician, in the sense of the elected official, presents a more difficult problem. The extremely weak motives for accumulating information on the part of the voter means that they are unlikely to detect any inaccuracy in his public statements on general matters. (Statements on some matter of special concern to the voter who is a member of a pressure group will be discussed below.) In an opposed election, however, there will be someone who is motivated to check statements and publicly attack the politician for lying if those statements are wrong. One solution which has been widely used is to imply things without actually saying them. If this technique is skillfully used, and most politicians have had much practice, the voters may think that a definite statement on some subject has been made but the politician's opponent may not be able to find the statement in such specific form that he can charge dishonesty. This is, of course, only one of the reasons why the politician's public utterances are normally so definite in claims that he will be a better choice than his opponent and indefinite in the reasons why this should be so.

Still, straightforward lying may be very helpful to a politician. With the voters largely inattentive, they normally only learn things about the campaign that the candidates force on them by public relations techniques. With most of the voter's information coming from the output of the candidates, and with most of the voters and sources of opinion already committed before the campaign starts, a repetition of a false charge may be very effective. The decision of whether or not to lie will depend largely on an estimate of the efficiency of the public relations organizations of the two candidates. If I can be confident that the lie will be more widely spread than the fact that it is a lie, I should lie $[BLP > (1\text{-}L)C_rL_r]$. Another useful technique is to have the false charge spread by someone else. You cannot be accused of lying, but some of the mud will stick. One can go on in this vein almost forever, and politicians have, in fact, developed an almost infinite set of practices of this sort. The basic point, of course, is the extremely poor information of the average voter and his lack of motives for improving that information.

The pressure group voter, however, has somewhat stronger motives for being at least moderately informed (explained in the last chapter). Furthermore, he is apt to have more casual information on the specific subject of "pressure" and will find it relatively easy to get more. Thus, although the member of a pressure group is unlikely to be really well informed, he is much more informed, and much more interested in getting more information, than is the average voter. Thus, the likelihood of his detecting a false statement is greater and the dangers of lying more severe. The sensible politician will actually support most pressure groups rather than simply saying that he is doing so.

This analysis could be extended, without much difficulty, to the question of the rationality of keeping a promise if it looks as though that might be inconvenient. In general, the problem would be that violating a promise will make it harder to get people to depend upon your word in the future. In areas like the black market, gambling, or the politicians' promise in a log-rolling "deal" where legal enforcement is impossible, the reputation of the individual for strict performance is particularly important. The basic equation could also be complicated to permit consideration of other factors.

The fact that lies are being propagated also affects the behavior of consumers of information. It means that it is harder to improve the accuracy of their judgment, the "A" in the equations in the last chapter, because some of the "information" they examine may have been designed to mislead. In essence, the cost in research of a unit increase in "A" is now higher, and normal economic principles would indicate that the consumer of information, if he is deliberately trying to improve some decision he is to take, will now choose to "purchase" less accuracy. Whether the existence of lies in the body of available "information" will lead him to put more or less time into research, will depend upon the elasticity of his demand for "A." The situation is exactly analogous to the increase in the price of an economic good.

Most consumers of political information, however, are not engaged in a rational process of trying to improve the accuracy of their voting decision, but simply picking up the information because they get satisfaction out of it. If the invented information was, somehow, more to their taste than the truth, they would be motivated to increase their consumption of it. Since it may be easier to tailor lies to the taste of the consuming public than it is to perform the same task with the truth, the introduction of entertaining but untrue stories about politics into the media may actually increase their circulation. In general, people are more strongly motivated to obtain correct information about items they are contemplating purchasing in the market, with the result that intriguing lies would have less attraction there. The existence of special services providing information for consumers of economic goods, while there is no *Consumers Digest* selling information to the voters on individual candidates, would be explicable on the basis of this consideration. The fact that we have rather stringent laws against fraudulent claims in the economic market place, while lies are not legally penalized in politics, might also be explained in this way. In the latter case, of course, there are many other possible explanations for the difference.

CHAPTER X

Proportional Representation

Democratic voting systems in use today may, in general, be divided into two major categories, the single member district system used in most English-speaking countries and a few other countries, and a system in which more than one member is elected from each district, with the seats in parliament being divided roughly in proportion to the number of votes each party receives. The Anglo-Saxon system is much the older, in fact the "proportional representation" (PR) system was invented in the late nineteenth century to eliminate some of its defects. The advantages of this newer method of representation are so obvious that most countries which have given serious attention to a choice between the two methods have chosen it. Nevertheless, the traditional methods of voting are so firmly established in the public mind in Anglo-Saxon countries that few citizens of these countries have even heard of this recent advance.

The original inventors of PR were concerned with the fact that the single-member district left a good many voters, those who had voted against the incumbent, unrepresented in the legislative assembly. In discussing this point with American political scientists, I have found that they use the word "represent" in a different way than did the European scholars who invented PR. To the American student of these matters, "represent" means little more than "elected from the district." Thus, they will say that a left-wing Democratic member of the House of Representatives "represents" a right-wing Republican who hates him and will never, under any circumstances, vote for him. In the European usage, "representation" is closer to an agency. A man only represents me in the legislature if, to some extent, he is dependent upon my support and/or has my approval. Presumably the left-wing Democrat would seldom, if ever, take the opinion of the right-wing Republican into account in pursuing his legislative duties. (He might even dislike the right-wing Republican enough to take a positive pleasure in doing things which will annoy him.) The Europeans wanted to make sure that minorities, even quite small minorities, were properly represented in the legislature by people of their choice, not of the majority's choice. That the legislature should more or less mirror the population was their proclaimed goal.

Before discussing the advantages of the system further, however, let us examine the possible institutional arrangements which have been tried or of which use could be made. First, I would like to discuss a system which I have invented myself, and which has never been used or, so far as I know, even been suggested. Indeed, until the development of the computer, the system would have been impracticable, which is probably the reason it has not heretofore been proposed. Let us assume that each representative in Congress simply be authorized to cast as many votes as the voters have cast for him. The total would be added by computers and the differential weighing of the various members of the representative assembly would represent their relative standing with the voters. No one would be unrepresented (with certain minor exceptions to be noted below) or "represented" by a man he detests. The voting on each individual measure would come as close to a national referendum as any representative body can achieve.

Note that this procedure would probably be more convenient if we combined our present districts into large areas, equal to five or more of the present districts.[1] This would permit popular members to acquire large numbers of votes, which would more accurately represent the popular will than artificially restricting them by geographical boundaries. In the extreme, the entire nation could be one voting district, with people running for Congress being free to pitch their sales appeal either to geographical areas or to groups of voters united by some nongeographical tie. It might be wise to make sure the result was not an elected dictator, however, by providing enough "constituencies" so that the most votes any representative could have would be some small fraction of the whole, say, 10 percent. Probably, however, this would be unnecessary. It is likely that the voters would spread their votes widely in order to get representation of their own particular political interests.

This system provides for electing representatives. It does not necessarily provide for defeating any candidate. With modern electronics there is no necessity for all representatives to meet in the same hall, consequently there is no maximum on the number of representatives. Voting could easily be done by wire, and the proceedings could be broadcast. In the extreme case, there seems no reason why people who wish should not vote for themselves and then fill their days by casting their single vote for and against the various proposals. Probably many elderly people and invalids would take advantage of this opportunity to obtain a feeling of importance and achievement. The pay of representatives, of course, would vary with the number of voters who had voted for them, and this would tend to lead candidates who had received few votes to decide not to spend their time on such an unremunerative occupation.[2] If it were desired, of course, a minimum number of votes necessary to be "elected" could be provided. This would keep the number of representatives down to some specified maximum if that is thought sensible.

So far I have not said anything about how this highly representative body would debate. I would like to defer any discussion of this problem until later since my suggestions are for fairly general rules which would fit many types of assembly. For the nonce, let us assume that only those representatives who receive more than some specified portion of the vote will be permitted to speak and sit on committees, with the remainder merely listening to the proceedings and casting their votes. Thus, the assembly would be perfectly representative of its constituents in its votes, but not in its debates, although the difference might not be much.

Real world PR systems are normally much less radical than the one I have outlined above, but they can be taken as efforts to approximate the same results without the benefits of computers. If each representative in the assembly has exactly the same vote as any other, then there will necessarily be a less perfect fit to the voter's desires, but the defect may be slight. The simplest system was proposed by Lewis Carroll. He suggested that if a candidate received more popular votes than the minimum necessary to elect him, he be permitted to give them to some other candidate. This would, over time, mean that popular candidates would have their effective vote in the legislature increased by the use of stooges. Interestingly, the system in use in the Netherlands is in legal theory very similar to Carroll's proposal.

If we are interested in representing the people in an assembly, however, there is no obvious reason why we should insist that this be done by individuals rather than corporate groups. Most democratic systems do develop such corporate groups in the form of political parties, and the voters fairly frequently are more interested in which of these parties "wins" than in the individual politicians who sit in the legislature. It is, of course, relatively easy to divide the votes in a legislature between several corporate bodies so that each has about the same proportion of the legislature as it does of the popular vote. In the simplest scheme, used in Israel, Sweden, the Netherlands, Austria, etc., the party decides who will occupy the seats assigned to it (normally by use of a priority list which is put on each ballot) so that the legislative representative is entirely dependent on the party machine for his seat. In a recent instance, a rather high official of the Mapai party of Israel went to the United States on government business and discovered, upon his return to Israel, that he had been moved far down the list so that he had no chance of being reelected to parliament. This power of the party machine over its representatives in the assembly is frequently regarded as a defect of the system. It is, if the members of the assembly are supposed to make up their own minds, but if we think of the voters as simply favoring one corporate body over the others, there is no reason to object to the corporate body casting its votes as an entity. In any event, as will be explained below, this is not an inevitable result of proportional representation. Voters can both select their party and the particular people who will represent it in the assembly.

Before discussing this possibility, however, let us turn to certain other methods which produce a legislature which is something like "a mirror of the voters." The first of these systems is used, so far as I know, only to elect a portion of the upper house in Japan. There are one hundred seats available in the "national constituency," a vast number of people run, and each voter casts one vote for his favorite. The one hundred who have the largest numbers are elected. Needless to say, this does not result in each voter being represented by a member of the legislature for whom he voted, and does not result in each member being elected by the same number of votes, but it probably comes closer to at least the first of these objectives than the Anglo-Saxon system of one-member districts. It also provides representations for "interests" in a particularly direct way. In this respect it should appeal to those who rather approve of functional representation à la the corporate state.

A second way of approximating proportional representation has been used, off and on, by France since 1870. Under this system, individual constituencies send single members to the parliament. It differs from the Anglo-Saxon system simply in that if no candidate gets a majority in the first balloting, a runoff is held in a few weeks in which only a plurality is necessary for election. These simple rules, together with the French talent for intrigue, have led to a functioning proportional representation system. A fairly large number of parties put up candidates in each constituency. Usually no single candidate gets a majority in the first round, and the period before the runoff is occupied by elaborate bargaining in which some candidates withdraw in favor of others. For example, the Socialist candidate will withdraw in one district and the radical in another, thus assuring the radicals of the first seat and the Socialists of the other. The system puts a high premium on ability in intrigue, and does not guarantee proportional representation, but on the whole has produced something very much like it.

There are two other systems which have been discussed theoretically but only rarely applied. In one of these, the voter has as many votes as there are seats to be filled in a multimember constituency, and can cast them in any way he wishes, all for one man, for example. My home state of Illinois uses this procedure for electing the lower house of the state legislature. A second system is the "single transferable vote" method which is what is described in most American textbooks as proportional representation. The system is complicated, little used, and already well described, so I shall waste no space on it here. Perhaps I should warn the reader, however, that I personally dislike this system, which seems to have the disintegration of parties as almost its only special effect. My decision not to discuss it at length, thus, may be an expression of prejudice.

I must now redeem my promise and explain how proportional representation of corporate bodies may be combined with selection of the actual persons who sit in parliament for the parties. In a sense this means combining the function of an American primary with that of an election. The simplest system is that in use in Italy. If you vote for the Communist party, you also are permitted to mark up to five individual Communists. After the total number of votes has been counted and it has been determined how many Communists will be sent to Parliament, the votes cast for individual members will be used to determine who will occupy the seats. A somewhat more elegant system is in use in Switzerland. Although the following description is greatly simplified, it is not, however, oversimplified. The voter may simply cast his vote for, say, the Socialists, and if so, he will help to determine how many seats the party will have in parliament. If he is concerned with who actually sits in parliament for the Socialists, however, he will strike the name of one or more of the Socialist candidates for office from his ballot and write in the name of some other candidate in their place. Since the man whose name he writes in is already on the ballot, this means he has voted twice for him and cast a sort of negative vote against the man he struck out. Only about 10 percent of the voters avail themselves of this privilege, and this minority of the voters determine who actually sits in the legislature. This procedure permits the election not only to sum the preferences of the voters but to give heavier weights to those voters who have more intense preferences.

Having rather sketchily surveyed the institutions available for proportional representation, let us now give somewhat more serious consideration to the reasons which may be urged for and against this form of representation. So far as I know, there are only three lines of reasoning used by proponents of the single-member constituency. Two of these, I am prepared to argue, are based on a misunderstanding of the nature of government. The reader should, of course, take warning; realize that I have an ax to grind, and that it is therefore possible that I am being less than fair to the system which is traditional in the Anglo-Saxon world.

The first standard argument for the single-member constituency is that it leads to a two-party system and one-party governments. To this is often appended the subsidiary claim that it also results in a succession of governments with policies which differ little because they represent about the middle of the electorate's opinion. This last part of the argument must be admitted, both observationally and in theory, to be true. The coalitions which normally govern in systems which do not have only two parties, however, normally also follow approximately the policies favored by the median voter, and for exactly the same reason. It is, therefore, impossible to argue for (or against) the single-member system on this particular ground. As for the basic claim, that the single-member system leads to a two-party system, the tendency is much weaker than normally assumed. Canada, for example, has a single-member district method of electing members of her parliament, but has an extraordinarily diverse collection of parties. Some of these parties represent distinct geographical areas, the French and English provinces, but in other cases they are not to be explained by ethnic differences.

It is fairly easy to design a model where the single-member

district will lead to a stable set of three, four, or five parties if there are geographical differences in the population. The Canadian system is an obvious example, the Parliament of Great Britain when Ireland was part of it, is another, and the American Congress is a third, if one is willing to recognize the distinctive nature of the Southern representation in Congress. In each of these cases we have multiparty systems essentially because of geographical differences. Since some of these systems developed out of a two-party system, as in Canada, it is clear that there is no normal historical progress toward two parties. It even seems possible that, where there are strong geographical differences within a nation, the proportional representation system may lead to fewer parties than the single-member district.

But even where there is no particular geographical component in the preferences of the voters, the tendency of the single-member district to lead to a two-party system is far from established. England, for example, is perhaps the best example of this system, and has had at least three parties represented in Parliament for more than sixty years. At present, one of the parties is much weaker than the other two, but the weak party of 1900 is now the government, which would suggest that there is no reason to believe that the smallest party will necessarily disappear. Further, for almost forty of the last sixty years, England has been governed by coalitions of two parties rather than by one party. Since this situation developed after many years of a two-party system, we again have evidence that the single-member district does not necessarily lead to a two-party system. The development of a new third party in Australia would also argue against the stability of the two-party system in single-member districts. (Australia, of course, has a transferable vote, but there is no obvious reason why this should affect the matter.)

If we examine more recent applications of the single-member district, the results are equally hard to fit into the two-party theory. In India there are many parties, but one wins all of the elections. The single-member constituency simply means that for twenty years the Congress party held three-fourths of the legislature, with about 45 percent of the popular vote, because its opposition was so splintered. The Korean case, although more complicated, is equally hard to explain under the two-party rubric. Indeed, it appears that the United States is nearly a unique example of a two-party system, and in order to use even that, it is necessary to ignore the difference between the Democratic party in the South and the Democratic party in the rest of the country. Historically, the United States has tended to move from a two-party system to a single-party system, which is followed by a splitting of the single-party into two. Whether this represents mere coincidence or is a natural consequence of the American electoral system, I do not know. In any event, the American electoral system is radically different from most others because of the direct election of the President and the primaries.

Even if, however, we agree that the single-member constituency leads to a two-party system, the reasons why this is an advantage are hard to discern. It is true that we are accustomed to this system, except in the South and New York, but tradition is not, by itself, a major argument. Neither the South, with its traditional system of primary elections nor New York City, with its more recent tradition of coalition governments, are examples of outstandingly efficient government, but they are not clearly inferior to the two-party systems by which they are surrounded. The normal argument for the two-party system is that it produces a single-party government which is "strong," as opposed to a coalition government. Granted that if you have a two-party system, one is likely to win the election, it is still true that the United States Constitution was built on the desire to keep the government divided. A majority of both houses of Congress and the President would all have to come from the same party (something which is almost unknown when the existence of the Southern bloc is taken into account) for such a "strong" government to exist. It has, in fact, been urged by prominent political scientists that the Democratic party which elects Congress is a different party from the Democratic party which elects the President.[3] Since we have gotten along for some time now with this system, it is hard to offer any practical evidence that it is a bad thing. It is possible to offer theoretical arguments for a unitary government, but I notice that people who make such arguments seem to aim at unitary government by the group which they think is closest to their own aims. They seldom argue for a strong unitary government when it seems likely that such a government would be run by their political opponents.[4] The system of checks and balances may not be the ideal government, but clearly it is not something which can be dismissed with a wave of the hand. People favoring a strong single-party government should explain why they really think they would be better off if the party which they least like had complete power in the way that the British cabinet has.

But in view of the difficulty of establishing that the single-member constituency leads to two parties, the matter can be left in abeyance. There is a third, much more subtle defense of the two-party system invented by Anthony Downs. He points out that a multiparty system will usually lead to government by coalitions of parties, and he points out that the voter can hardly know in advance what coalition will form, and what policies it will follow. Thus, if we accept this argument, the multiparty system suffers from the very serious defect that the voters do not know what they are voting on. Clearly, this could be a very serious defect. Perfect knowledge, the voters will never have, as many people who voted for Johnson are beginning to realize, but we do have some idea of the policies to be implemented by the party we choose if it wins.

Obviously, this is an argument which deserves careful consideration, and I will not be able to disprove it in any strict sense. I will, however, suggest that it puts the emphasis on the wrong point. We are interested in controlling the government, not strictly speaking in the knowledge we have when we cast a vote. The deficiency in knowledge pointed to by Dr. Downs seems to also be a deficiency in our possibility of control, but this has not been demonstrated. Suppose we contrast a system of five parties, generally representing five different political positions with two parties. First, the information transmitted to the politicians by the votes in the five-party system is obviously much greater. This information differential does not result solely from the fact that the voter can cast his vote five ways instead of two, and hence can more closely approximate his opinion in his vote. It also results from the different conditions of the parties themselves. Downs proved that in a two-party system the parties will have very similar platforms, and our empirical evidence would seem to indicate that this is so. In earlier chapters, I have demonstrated that in multiparty systems the parties are well advised to keep some distance apart in their positions. Again, the empirical evidence seems to support theory. Thus, the votes transmit much more information under the multiparty system than under the two-party system. The shift of votes from one to another of a pair of virtually identical parties with virtually identical platforms shows little about the opinion of the voters. Shifts among a number of quite different parties are much more informative.

Dr. Downs's answer to this presumably would be that the voters would not be able to make sensible initial choices among the many parties because they would not know what the shape of the eventual coalition government would be. Thus, he would argue, very little information would be transmitted by the votes because the voters would cast them in ignorance. This, of course, assumes that the voters must vote for a government, not for a sort of agent who will negotiate to form a government. I may have no real idea what parties will make up the coalition government after the election, but still be very much interested in strengthening the bargaining power of a party which is urging further aid to agriculture. If I vote for the agricultural party, I can be no more confident that the party of my choice will eventually be part of the coalition government than the voter in the two-party system can be sure that the party he votes for will win. I can, however, strengthen my party, which will improve its negotiating position and hence improve the likelihood that the policies I favor will be a part of the government's policy.[5]

There, thus, does not appear to be any decisive solution to the problem raised by Dr. Downs. Still, if there is a reasonable argument for casting a vote for the party which most closely approximates your own opinion in a multiparty system, then it is likely that the voter will do so. The fact that some political scientists doubt that he can really make up his mind intelligently will not prevent him from trying, and if he tries, the information content of his vote will be the same as if he were fully informed about the outcome of the coalition negotiating process. Thus it is possible, although not certain, that the objection raised by Dr. Downs is largely irrelevant; that the individual does not need to know the policy which will be followed by the government coalition formed after the election to cast an informed vote. Even if this possibility were not present, however, the improved information content of the vote itself to the politicians should provide a counterbalance to the difficulty he has emphasized. Still, the issue is a difficult one, and I would not like to give the impression that I feel that Dr. Downs's objection to the multiparty system is entirely without merit. Clearly, this is a problem which will repay further study.

So far I have talked about the arguments for the single-member district system, and found little merit in them. Now let us consider the positive arguments for the proportional representation system. Before turning to the major arguments, however, a minor point should be made. Proportional representation can, and has produced single-party majority governments. For many years Norway, Denmark, and Sweden were governed by powerful single parties who in each case held a majority of the seats in the legislature. If one party succeeds in convincing a majority of the population that it can best represent them, then it will be able to elect enough members to dominate the legislature (this is not necessarily true of single-member constituency systems). Thus, coalition governments are not inevitable under proportional representation. They occur only when the voters do not give majority support to one party. Nevertheless, when the voters are given more than two choices, they quite frequently do not give any one party a majority, with the result that some sort of coalition is formed to govern the country. Since our own system normally leads to a coalition of the Southern Democrats and one or the other of the Northern parties dominating our legislature, this should not impress Americans as a bizarre phenomenon.

Turning now to the arguments for proportional representation, the original argument for it has been rehearsed above. It aimed at insuring voters of representation in the legislatures for whom they had voted. The inventors of the system in the nineteenth century thought the single-member constituency system, which was already old at that time, deprived the minority in each district of any real say in the legislative assembly. They were not content with believing them to be represented by the man against whom they had voted, and whom they might well distrust to a great degree. This seems to me a respectable motive, but I would like to refer people who wish to see it elaborated to the nineteenth-century literature on the subject, and devote the rest of my space to some newer arguments for the system.

The first of these newer arguments may seem much more economic than political. Economists regard a system of competition between two companies as better than a monopoly, but far from ideal. In general, it is too easy for the two competing corporations to find a common bond in exploiting the public, to become conservative and shun innovations. The more competitors, the more pressure on each one to behave in an efficient manner. The details of this argument can hardly be carried over to politics, but its general outline applies. In particular, the two-party system bars new parties from entering if the two existing parties get out of touch with the voters' desires. The multiparty system lets them in easily. Further, the party out of power in a two-party system is likely to have had at least some responsibility for the initiating of many policies now being implemented by their rivals. Under the circumstances they may have good reason not to engage in violent criticism when a third party would not be so inhibited. It is much harder to keep five parties, together with some small splinters who would like to be parties, quiet on a matter in which the government is in difficulties than it is to keep a single opposition party satisfied to leave the matter unmentioned. In this sense, the voter is apt to get much better information from a multiparty system. Sacred cows are much safer in an atmosphere in which there are only two parties, each trying to appeal to the middle range of voters, than if there are five, each appealing to a rather different audience.

More importantly, however, the proportional representation system has the advantage that it insures that a majority in the legislature represents a majority among the voters. The single-member constituency system permits a majority in a majority of the districts, in the limit 25+ percent of the population to control the government. It seems likely that log-rolling normally goes on under roughly these conditions, although a two-house legislature can somewhat mitigate these problems. If this is so, then the results of voting must be highly nonoptimal.[6] The switch to proportional representation would result in the minimum coalition of voters which can get a measure through being raised to 50+ percent. This may not be optimal, but it is far better than 25+ percent.

The main purpose of this essay, however, is not to prove the superiority of proportional representation, but simply to acquaint the reader with the range of alternatives open. The first version, my own invention, seems to me the best in an age of computers, but the others approximate it to varying degrees of accuracy. It could be easily adjusted to permit corporate bodies (usually called parties) to be represented, and could also be made to permit the voter to not only select the corporate bodies but also decide who shall be on their boards of directors. It would even be quite simple to make the system follow the Swiss in giving the more intense voters greater weight. Instead of elaborating these rather tedious details, however, I would like to close with a brief examination of a problem of representation which is hardly ever seriously discussed: the allocation of time during the debates.

One of the most brilliant, and at the same time one of the most ignored, pieces in political theory was Bertrand de Jouvenel's "The Chairman's Problem"[7] in which it was demonstrated that most members of a legislative body (or public meeting) can play little part in the debates. His classically simple way of showing this was simply to divide the time to be devoted to a given subject by the number of members to get the average time available for each member if each could speak. The result is normally a period too short for effective presentation of a point of view. De Jouvenel then suggested that what actually happens is that the presiding officer selects a few persons who he thinks will more or less cover the spectrum of opinion, and lets them speak. The number of speakers would not, if the theory is true, vary much with the size of the legislative body. This, of course, is in accord with empirical observation, although it is not at all certain that the selection of the speakers is always through the mechanism suggested by De Jouvenel.[8]

In practice, methods of managing debate vary tremendously. The English Parliament, for example, has a well-worked out system under which most members are seen but not heard very much. They aren't even seen as much as might seem normal since the House does not have enough space for them all to sit down at once, with the result that some of them are necessarily always absent. The House of Representatives, by contrast, makes a conscientious effort to give every member a fair share of its "debating" time. As a result the normal situation in the House is an almost empty room[9] in which an obscure representative is droning on about some subject which interests him. Most representatives turn up for votes, and important debates will attract a sizable crowd. In the important debates, also, the average member has little chance of being recognized if he is so temerarious as to ask for the floor. The time is largely taken up with the leaders of the House, almost as if it were the British Parliament.

This general system of selecting people has worked reasonably well for thousands of years, and there would appear to be little reason to change it. Here again, however, the invention of the computer makes it possible to do better. Let us assume that we have a legislature of one hundred persons. Each one has a "right" to one one-hundredth of the time. The computer could easily keep track, crediting each person with a point for each moment of listening, and subtracting one hundred points for each moment of speech. The individual members would be permitted to either hoard their own points or to give (sell) them to others. Thus, each point of view would have an allotment of time strictly proportional to its strength among the voters.[10] The computer, in fact, could take over many of the more routine duties of the presiding officer. It could, for example, choose the next speaker by simply noting how many points each applicant had and giving priority to the one with the most.[11] If a number of members wished to speak on a given subject, and they did not expect the computer to select them in the order they desired, they could make use of the present "yielding" system to order their arguments as they desired. The presiding officer could, being relieved of the necessity of deciding who was to speak, devote himself to the more complex parliamentary problems which periodically come up. Individual members would not be subject to control by the speaker, because any small group which felt oppressed could always arrange to pool their points and get a chance to put one of their members into any debate they wished.

It may appear that this is a wildly unconventional suggestion. It is. The view that the procedure we have become used to over the last 2000 years is necessarily the best, however, seems highly suspect. Only by considering new ideas can we hope to improve, and only by reconsidering even the best established institutions can we decide where improvements would be desirable. The whole point of the "new political science" is to raise questions about received doctrine, and to try to find the best answers to both the traditional problems of politics and the new ones. New ideas always seem radical and bizarre. I would not claim that the new ideas I have advanced in these essays are the best possible suggestions. I hope, however, that they will play at least some role in the search for a better and more scientific political structure.

Notes

Notes to Chapter I

1. There has been a good deal of economic research concerned with advertising in the last few years. "Advertising and Competition," by L. G. Telser (*Journal of Political Economy*, December 1964, p. 537) will do as an example. The bulk of this research has dealt with advertising as a source of information. I do not, of course, deny that advertising transmits information, but it may also change preferences. It is with this aspect of advertising that this chapter is concerned.
2. Madison Avenue, of course, is called in to help. Not only are most political campaigns now guided by experts from the commercial advertising world but the government frequently solicits their assistance to implement such policies as selling savings bonds and trying to convince people they should not smoke.
3. "Compulsion, Where's the Compulsion? You have a perfectly free choice. You can sign that paper or be hung," Captain Peter Blood.
4. It will be even more similar to a model recently published by Kelvin Lancaster ("Change and Innovation in the Technology of Consumption," *American Economic Review*, May 1966, p. 14; and "New Approach to Consumer Theory," *Journal of Political Economy*, April 1966, p. 132). The two models were derived independently, but a careful examination of their resemblances and differences seems necessary. This discussion will be deferred until my model has been fully presented.
5. Due to the possibility of indifference, the full proof is rather complicated. See my "The Irrationality of Intransitivity," *Oxford Economic Papers* (October 1964), p. 401.
6. This was, I believe, the majority view of the matter, although there was a vigorous minority on the other side. For a more comprehensive presentation of my former view, together with footnote citations of most of the relevant literature, see "The Irrationality of Intransitivity," *op. cit.*.
7. Suppose we consider the establishment of a new "service" industry

which gives its customers floggings in return for a reasonable fee. It will be generally agreed that no matter how zealously the slogan "Fifty Lashes Every Friday" was promulgated, the company would be unable to establish a mass market.

8. I have drawn the line "single-peaked." This is in conformity with my personal opinion that single peaks or single troughs would normally characterize such curves. This opinion, however, is based on the situation with several variables and it seems sensible to delay discussion of the point until we discuss that situation.
9. If there is a difference in cost, then the situation will require the multidimensional geometry to be developed later.
10. Duncan Black, *The Theory of Committees and Elections* (Cambridge, 1958).
11. *Op. cit.*
12. Since Lancaster's model was published first, it might be better to say that mine is close to a mapping of his. However, the two models were developed completely independently, and the use to which Lancaster puts his is rather different from that to which I put mine.
13. This elegance has some drawbacks. In order to make his system mathematically simple, Lancaster has implicitly assumed that increases in income do not change the assortment of goods purchased. This can easily be changed on his geometrical diagrams by substituting curved lines for his straight rays, but this would raise difficulties for his algebra.
14. Note that the simple, single mountain preference pattern would not always be found. If, for example, we put the color spectrum on the horizontal axis and the weave of some cloth on the vertical axis, the preference surface might be extremely irregular. Where preferences are like this, the social organization must be capable of producing a very large number of different "products." The present organization of the women's wear trade, with its ability to produce a staggering variety of goods in small batches, is an example.
15. These critics have, for some reason, confined themselves to commercial advertising, and tended to ignore the work of political and governmental advertisers.
16. Democracies normally also have people who feel that governmental and social activities designed to make other people's tastes conform more closely to their own are desirable. Since these people tend to be both intellectually and socially prominent, they do get some action of this sort organized. They may even have some effect. Usually, however, in a free society someone else is providing the prospective "beneficiary" of these activities with an alternative. BBC's efforts to uplift the common people of Britain made a lot of money for the stockholders of "pirate" radio stations.

Notes to Chapter II

1. This may be interpreted as a "patronizing" on my part of political scientists, but this is not the case. We live in a world of speciali-

zation and many experts in one field know little about the techniques used in other areas. Indeed, granted the limited capacity of the individual human brain, it would be impossible for any scholar to be fully conversant with all the methods used in other fields.

2. In a space where all points are not available, our discontinuous apples and oranges, for example, the exact tangency conditions may not be met, and the locus of all points from which no change will result, may not be quite so regular as drawn. Fortunately this makes little difference, and we can use the line as a close approximation even in these cases.
3. Professor Oliver Williamson has suggested that the individuals might want all of the apples and oranges even though they were planning on making a redistribution after they received them. Thus, the avoidance of starvation would not be relevant. My way of diagramming the situation, however, is surely possible, and provides an introduction to further analysis.
4. *International Economic Papers,* No. 9, pp. 39-92.
5. I have drawn these figures so that all of the optima are either on the exterior of the figure or very close to the exterior, as in the case of C in Figure XII. This is not a necessary characteristic, and different arrangements will be briefly discussed.
6. A line, strictly speaking, has no width. Hence, in order to speak of the area occupied by the Pareto points as compared to the whole area, it is necessary to think of the issue space as being discontinuous instead of continuous. If the space shown on Figure VII consists of 10,000 points and the line on Figure V consists of 100 points, then the percentage of the points occupied by the Pareto region in Figure VII is clearly very much smaller than the percentage occupied by the Pareto region on Figure V.
7. The University of Michigan Press, 1962.
8. If the injuries and benefits from individual decisions were apt to be very great we might wish to buy insurance by adopting a rule which gives both less variance and a lower "profit."
9. This assumes that cycling does not occur. My reasons for doubting that cycling is an important phenomenon will be presented in the next chapter.
10. The people injured may be injured more than the people benefited are benefited, of course; this is one of the reasons why simple majority voting is unlikely to be optimal.
11. Kenneth Arrow, *Social Choice and Individual Values* (New York: John Wiley & Sons, Inc., 1951) pp. 48-51.
12. Or by ordinary bargaining. In a two-person world, both require unanimous consent, hence they are identical.
13. Preference functions in which there are two maxima for an individual, the mountain rises to a flat plateau instead of a peak, etc., are possible, but these things are more curiosities than real problems.
14. See *The Calculus of Consent,* and "Public and Private Interaction Under Reciprocal Externality," by James Buchanan and Gordon

Tullock. *The Public Economy of Urban Communities, Resources for the Future* (Washington, D.C., 1965), p. 52 and especially p. 63.

15. This use of this sort of space has been developed by Duncan Black, *Theory of Committees and Elections* (Cambridge, 1958). See also Duncan Black and R. W. Newing, *Committee Decisions with Complementarity* (London: William Hodge, 1951).
16. If the three optima are thought of as on the contract loci, then this statement must be modified since it is possible for the optima to be excluded from the cycles. In fact, at least one of them is certain to be.
17. See *The Calculus of Consent*, pp. 148-50; and John Von Neumann and Oskar Morgenstern, *The Theory of Games and Economic Behavior* (New York: John Wiley & Sons Inc., 1944), pp. 222-30, and 264.
18. For simplicity in drafting, it is assumed that all indifference curves are perfect circles. In the next chapter a similar assumption will be made with a more serious purpose.

Notes to Chapter III

1. I regret to say that the phantom has stalked my classrooms with particular vigor. I hereby apologize to my students for inflicting it upon them.
2. The situation with complete independence of preferences is discussed in "A Measure of the Importance of Cyclical Majorities," by C. D. Campbell and G. Tullock, *The Economic Journal* (December 1965), pp. 853-57. Since publication of this note a number of other scholars have made significant contributions to the subject. Unfortunately, the most important of these have not yet been published, but are circulating in mimeographed form. Mark Pauly of Northwestern University is preparing a book of readings which will contain all of the work in the field, and the interested reader, if he will have a little patience, will find the material there.
3. John Wiley & Sons, Inc., New York, Revised Edition, 1963.
4. Pp. 74-91.
5. For fuller discussion of single-peaked curves, see Duncan Black, "On the Rationalle of Group Decision-Making," *Journal of Political Economy* (February 1948), pp. 23-24, and *The Theory of Committees and Elections* (Cambridge, 1958).
6. London: William Hodge, 1951.
7. Pp. 51-59.
8. Usually the limitations on the introduction of trivial amendments is not this formal, but it normally will be impossible to make very small changes in money bills.
9. Strictly speaking there would be an indifference point when the perpendicular bisector ran directly through the point of maximum preference for one voter. At this point the vote would stand: for A, 499,999; for the alternative, 499,999; indifferent, 1.

10. A reader of the manuscript suggested that the last two paragraphs be replaced by the following passage: "Suppose that two points, A and B, are equidistant from the center but that *no one's optimum lies exactly on their perpendicular bisector.* Therefore one, say A, is preferred to the other. All points within a small neighborhood of A are preferred to B, and some among these will be further from the center than B is." A little experimentation on the understandability of both my version and his seems to show that some people find one easier to follow and some the other. Including both, therefore, seems sensible.
11. If there were some group or person who had control of the order in which proposals were offered for vote, and if that group or person had perfect knowledge of the preferences of the voters, then by proper choice of alternatives it would be possible for the "Rules Committee" to arrange the vote in such a way as to lead to substantially any outcome it wished. See Benjamin Ward, "Majority Rule and Allocation," *Journal of Conflict Resolution,* 1961, pp. 379-89.
12. For a proof of the normal absence of a majority motion under assumptions like those we are here using, see Duncan Black and R. W. Newing, *Committee Decisions with Complementary Valuation* (London: William Hodge, 1951), pp. 21-23.
13. *The Theory of Committees and Elections* (Cambridge, 1958), p. 18.
14. Occasional cases which appear to involve cycles have been discovered. See William Riker, "Arrow's Theorem and Some Examples of the Paradox of Voting," in S. Sidney Ulmer, Harold Guetzkow, and William Riker, *Mathematical Applications in Political Science* (Arnold Foundation, Southern Methodist University, Dallas, 1965), pp. 41-60.
15. Most rules of procedure have such provisions, but they are usually easy to avoid if a majority favors the measure which is reintroduced.
16. If they differ in only one characteristic, then the work of Duncan Black indicates that the cyclical majority is most unlikely.
17. Circles in two-dimensional issue space. When more than two variables are considered, the issue space would have more than two dimensions and the indifference hypersurfaces would be hyperspheres.
18. In a very kind letter Professor Arrow agreed with the argument presented in this chapter. However, he pointed out that the following portion is not mathematically strict. He expressed a desire for "a stronger and stricter statement." I also would like to convert what is now a strong argument into a mathematical proof, but have been unable to do so. Perhaps some reader will be able to repair the deficiency.
19. Even with large numbers of individuals a combination of certain strong symmetry relations between their preference curves and special arrangements of their optima will lead to significant cycles. The only cases I have been able to develop involve very

unrealistic shapes for the indifference curves and there seems no point in presenting them here.

Notes to Chapter IV

1. Harper, 1957. Downs's contribution was not, of course, limited to application of Hotelling's model.
2. For Downs's conditions see *An Economic Theory of Democracy*, pp. 23-24.
3. This assumption differs from that used by Downs, but permits somewhat simpler reasoning. See Gordon Tullock, *The Politics of Bureaucracy* (Public Affairs Press, 1965), pp. 88-96. The strict Downs model will be generalized later.
4. *Ibid.*
5. In *The Politics of Bureaucracy*, the voters are distributed, for simplicity, evenly along the line. If they are projected from the space onto the line, they would only, by coincidence, be so distributed. Fortunately, all of the conclusions drawn from the linear model are true regardless of the distribution.
6. Yale, 1962.
7. John Moore, who read the manuscript of this book, suggested that the problem here is closely similar to the difference between a "sales maximizing" and a "profit maximizing" model of the firm in economics.
8. This conclusion is, of course, Riker's. This whole section is a translation of Riker's reasoning into our geometrical model.
9. An example using a circular issue space may be more helpful to some. Suppose the circle is 1000 inches in radius and we start with three parties at the points of an equilateral triangle, one inch on a side, and centered on the exact middle of the circle. The perpendicular bisectors of the three sides of the triangle will divide the circle into three 120 degree pie-shaped sections. Suppose one of the parties moves directly outward .52 inches. This changes the triangle to a 40 degree, 70 degree, 70 degree shape and gives the party which had moved a segment of 160 degrees. This segment will have its apex .45 inches away from the center, but this loss will be much less than the gain from the widening of the angle.
10. In practical politics this degree of difference in intensity would not be likely unless there were a great number of other groups each interested in its own small issue.

Notes to Chapter V

1. The spatial model used closely resembles that introduced by August Losch in *The Economics of Location* (Yale, 1954), by Woglom and Stolper. See especially his Chapter 9, "The Market Area."
2. My "Optimality with Monopolistic Competition," *Western Economic Journal* (Fall 1965), p. 41, covered much the same ground as the next few pages.

3. Given the present assumptions, the reduction in price by the two adjacent drugstores would not affect the prices of the remaining drugstores. This is a result of the very thin distribution of such stores, however. None of the original stores are close enough together to compete directly. Once the stores are relatively close together, the introduction of one new store may affect prices in far distant locations by a sort of chain reaction. In a rather thin, one-dimensional system, game-theory considerations might come in, with the different stores trying to out-game each other, but in the real world, with a two-dimensional system and each store having more than two "neighbors," this would not be so.
4. With a denser set of drugstores, this change in prices would spread much farther, and hence the injury and benefit would be much more widespread.
5. The location of the end drugstores at the two extremes requires some special assumptions and elaborate reasoning. Since the equal distribution model is of little importance, we shall ignore the problem.
6. This point is developed at considerably greater length in my "Optimality with Monopolistic Competition," *op. cit.*
7. Strictly speaking this is only true if the total number of ice-cream cones sold is increased by the cut in price. Since this condition would surely be met, we need not worry about it.
8. Principally by the "Chicago School," of which I am a (dissident) member.
9. The cost curve could change just as rapidly as the cost of transportation.
10. An arbitrary selection of this point for the purpose of statistical testing would not be hard to make.
11. See Chapter I for a discussion of this sort of individual preference curve.

Notes to Chapter VI

1. There is some "demand" for such changes, however. Art appreciation courses are normally taken by people who wish to change their own taste.
2. The great importance put on "freedom of speech" by most democrats can be taken as an expression of the opinion that such things are not externalities. The opposite opinion of most totalitarians can be taken as upholding the other view. A difference in perspective is important here. The democrat is uninterested in the receipt by governmental personnel of wages which are above the opportunity costs of governmental employment. A totalitarian government will normally be mainly concerned with keeping its "consumer's surplus."
3. Some information is absorbed for strictly business reasons by people who would be doing something else if they didn't need the money. In certain areas this phenomenon is of considerable importance, and will be discussed below, but mostly the "customers" of the

information and entertainment industry are motivated by the direct satisfaction they get out of their consumption.

4. The principal exceptions to this rule concern information which is acquired "in the line of business." In total this information may be very great, but it is mainly transmitted through highly specialized channels rather than the publicly available media.
5. In order to avoid confusion, the word "information" is used consistently for all things which might affect opinions or tastes. This is something of a misuse of the word since lies, errors, and bursts of emotion with little factual content may all fall within our usage but would be considered outside of the ambit of the word in normal speech.
6. Due to the copyright laws each magazine, TV station, etc., actually has a pure legal monopoly on its product. These products, however, are highly intersubstitutable, and the result is not really monopolistic.
7. There are some limitations to this principle due to the fact that literature distributed at a zero price (junk mail, for example) may be disregarded just because it is so cheap. The low price is seen as an indication of unreliability.
8. The bell-shaped assumption is used for convenience. Actually, any assumption which differs from absolute evenness will do.
9. The congestion factors, which made the densely populated areas in our geographical model indeterminate, do not apply here.
10. Although this seems very concentrated, American consumers are lucky. In much of the world there is only one network.
11. In a way, there is actually a negative effect of such programs. It is likely that opera or other cultural programs not only attract fewer customers than do the more normal programs but the people they do attract are somewhat less open to influence by the commercials.
12. The distances are shorter unless the stations are so far apart that there is no overlap at all in their potential audiences. Given any concentration of the audience at all, this is most unlikely.
13. Note that I have only said this "may" happen. If the bell-shaped distribution of potential watchers held more people between A and Y than between A and X, and if certain other assumptions were made, the leftward movement of the people between A and Y might more than offset the rightward movement of people between A and X.
14. Robert Meister, "The McLuhan Follies," *The New Leader* (October 10, 1966), p. 20.
15. "My point is simply that human beings and other higher organisms need a mixture of order and novelty, of the familiar and the new, if they are to be reasonably satisfied with their environments." (New York: Holt, Rinehart & Winston, 1966), p. 61.
16. That this is not an irrational pattern of preferences may be seen if we consider that individuals cannot accurately judge the truth or falsity of new information unless it has a fairly close connection to what they already know. Scientific articles normally are heavily

footnoted to existing information in order to make the new contribution acceptable to the reader. Michael Polanyi has argued that a conservative attachment to a good deal of the existing orthodoxy is an indispensable condition for scientific progress. The individual, too, moves to new knowledge by a series of small steps rather than by giant leaps.

17. This may not be so much of a disadvantage. Attempting to guess what people will buy, from what they say they will buy, is normally a thankless task. See Robert Ferber, "Anticipations, Statistics and Consumer Behavior," *The American Statistician* (October 1966), p. 20.
18. From the standpoint of society as a whole, the activity of all "public relations" operations might still be random.
19. If it is inherently attractive to a sizable group, but not to most people, then specialized media will concentrate on it.
20. See my *The Organization of Inquiry* (Durham: Duke University Press, 1966) for a more comprehensive discussion of the process.
21. The advertisements, of course, make it possible to sell the journal for less, and this may be more attractive to the readers than the additional editorial material. Further, some advertisements actually attract readers (James M. Ferguson, *The Advertising Rate Structure in the Daily Newspaper Industry* [Englewood Cliffs, N.J.: Prentice-Hall, Inc., 1963]). Nevertheless, basically the advertisement is a deliberate attempt to "sell" an idea with full knowledge that it does not maximize circulation.
22. Unless, by coincidence, what they wanted to push was also the circulation-maximizing position.
23. See Oliver Williamson, *The Economics of Discretionary Behavior, Managerial Objectives in a Theory of the Firm* (Englewood Cliffs, N.J.: Prentice-Hall, Inc., 1963) for a general discussion of the deviations from profit-maximizing which are possible for utility-maximizing managements.
24. This sort of strategic move is useful only if it is confined to one group. If both the right and left were to purchase magazines strategically, they would largely cancel out, with the result that there would be a net social loss with no resulting shift in opinion.
25. Few in comparison to their total readership. *Time* and *Life* receive what I would call "many" letters even though they represent only a tiny fraction of the readers.

Notes to Chapter VII

1. Harpers (New York, 1958). In general I will not provide footnotes for each point where my analysis duplicates that of Downs, since there are a great many such places.
2. Since the voters in a primary are normally much less numerous than in the regular elections, they may be a select group, and better

informed than the average voter. This may partially counterbalance the effect of the poorer information available.

3. Again, this is only true as a general rule. Voting decisions may be made as the result of careful consideration of the issues. The switch of Iowa from firmly Republican to Democratic in the 1964 election fairly clearly resulted from the attitude of Barry Goldwater and the Republican congressional delegation on the continuation and expansion of farm subsidies. A good many firmly Republican farmers were forced to make a choice between their principles and their pocketbooks. They chose their pocketbooks, but clearly this involved a good deal of thought for most of them.
4. The action could be writing his congressman, or simply keeping the congressman's vote in mind until the next election.
5. Note that the cost would be largely in the form of leisure time devoted to study. Thus, the individual's relative appreciation for different ways of spending his leisure determines the cost of becoming informed. Here, again, the reader with both intellectual tastes and a good background must guard against assuming that the costs incurred by the average man would be as small as the cost to himself.
6. Log-rolling might lead to such a bill, passing as part of a much more complex chain of legislation. This is not strictly relevant to our present line of reasoning, but it should be noted that one of the by-products of log-rolling is a degree of complexity in the governmental process, which makes it considerably harder for the voter to evaluate any given proposal.
7. By James Buchanan and Gordon Tullock (Ann Arbor: The University of Michigan Press, 1962).
8. Although indirect taxes are normally thought of as concealed, this is only from the general public. In general, indirect taxes are very, very conspicuous to those individual members of the public who actually pay them. A manufacturer's tax on steel may be invisible to the average voter, but the steel manufacturer knows about it.
9. In the casual-information model, which will be discussed in the next chapter, further effects of voter ignorance will be developed. These additional effects are ruled out by our present set of assumptions.
10. Something rather like this does happen with local government units.
11. See Aaron Wildavsky, *The Politics of the Budgetary Process* (Boston: Little, Brown, and Company, 1964), for a discussion of the actual methods now used.
12. This equation was somewhat further developed and subjected to empirical tests in "A Theory of the Calculus of Voting," by William H. Riker and Peter C. Ordeshook (unpublished manuscript). The Riker-Ordeshook study which was, in part, the result of a reading of this book in draft, on the whole supports the purely theoretical conclusions drawn here.
13. There is a sort of concealed assumption that all votes are simple

choices between two alternatives. This is realistic if we are talking about American elections and is also much simpler than choosing one among a number of alternatives.

14. Dr. Anthony Downs must be listed among the people who are shocked. In fact he continues to vote regularly.
15. *Op. cit.*
16. In addition to these restrictions which are necessary to prevent corruption, it would be necessary to make sure that the purchaser of the vote used his own money. If the Ford Foundation, for example, were to provide subsidies, the whole experiment would be severely biased.
17. Subjectively, B is also possibly a function of information. Learning more might increase or decrease the voter's estimate of the importance of "his" side winning. Objectively, of course, the real benefit does not change.
18. This point was first made, I believe, by Dr. Roland McKean at the Faulker House Conference in October 1963.

Notes to Chapter VIII

1. It should not be forgotten that many voters do not even read newspapers. In the course of interviewing an applicant for a teaching appointment, who had a recent Ph.D. in economics, I discovered that his only source of news on the course of politics was radio newscasts.
2. Recent research indicates that the poor owe a good deal of their low standard of living, not to their low incomes, but to their inefficiency in purchasing. They buy shoddy merchandise at prices higher than they would need for better quality. D. Caplovitz, *The Poor Pay More* (Glencoe, Ill.: The Free Press, 1963).
3. Among intellectuals, a formal bow is normally made to impartiality by voting for a couple of minor candidates in the "other" party at each election.
4. As a limitation on this principle, there may be some voters who, for one reason or another, cannot affect the outcome of the election. Severe injury can safely be inflicted upon them.
5. By Gordon Tullock (Thomas Jefferson Center. Monograph No. 5, University of Virginia, 1962).
6. In a great many American cities a morning and an evening paper are jointly owned. It is perfectly normal for these two papers to take different editorial positions.
7. In the academic community, those members who are taking an active role in politics are apt to be quite well informed. They are, of course, a small and biased sample. A fairly short conversation with the people who exert low-level political leadership in nonacademic environments is enough to indicate that for them, C_i is low.
8. These costs are largely imposed by members of his own department. Occasionally, however, the central administration will get upset about what is being taught in some course and attempt disciplinary action. Normally the "victim" of this process will find numerous and vigorous allies and will suffer no net costs. The

situation is quite different if the junior member of the faculty differs with the members of his own department. Fortunately, many departments are quite tolerant of such deviations.

9. Needless to say, this effect may show up in a retardation of movement in an undesired direction rather than in actual movement in the desired direction. Any one individual will be only a tiny fraction of the influences in action at any time.
10. For simplicity I assume the same ten persons throughout. The same process may take place more slowly with different generations being involved. The usual book in intellectual history will take the form of an account of how some such idea as "E" gradually spread through a society over a considerable period of time.

    Our diagram can be regarded as simply a geometric representation of this sort of growth, although I have included some ideas that did not spread.
11. Strictly speaking, this involves the additional assumption that the holders of the other ideas do not have strong enough feelings to log-roll E down. During the 1964 election, 88 percent of the population favored permitting prayer in the schools. That clever politician Lyndon Johnson, however, realized that the minority which was opposed felt much more strongly on the issue, and hence that there were more votes to be gained by supporting the Supreme Court than by supporting prayer.
12. In Figure LXVII these "unpopular" ideas predominate in Period I, but there are two of them, C and F, still in existence in Period V. Ex ante it is no less likely in Period V that F will be the dominant idea in Period IX than it was in Period I that E would be the dominant idea in Period V.
13. Any new idea is, naturally, held by a tiny minority (originally a minority of one). In the natural sciences such ideas rapidly are tested by the majority of the profession. We need, but are unlikely to obtain, a similar mechanism for the social sciences.
14. See "'Realism' in Policy Espousal," Clarence E. Philbrook, *American Economic Review* (December 1953), p. 846.
15. This argument can, without much difficulty, be made stochastic, with different people having different "attention spans."

Notes to Chapter IX

1. This will be particularly likely if the parts of the propaganda which mention the material gain to the farmer present this gain as a method of obtaining a more general good. Thus, statements that agricultural prosperity is essential to the national prosperity not only provide the farmers with a rationalization for the program but also make it likely that the nonfarmer who happens to read about the program will not interpret it as simply aimed at getting money for farmers.
2. For a general discussion of the whole problem of the organization of pressure groups, see Mancur Olson, Jr., *The Logic of Collective Action* (Cambridge: Harvard University Press, 1965).

3. No Nobel prizes have been awarded, presumably partly because of the reputation of the scientists involved and partly because accident seems to have played a larger role in the discovery than it usually does.
4. Unless his belief is somehow communicated to the hearer, perhaps by tone of voice. If this occurs, however, the person attempting to lie has failed, since he has conveyed the truth to the listener although he did not intend to. An untrue statement expressed in such a way that the hearer realizes it is untrue may be a lie in the strict use of English, but it is an abortive, nonfunctional lie. For our purposes, we shall ignore this class of statements.
5. Ed. by Gordon Tullock (Columbia, S.C.: University of South Carolina Press, 1961), p. 96.
6. It might be believed long enough so that the desired action was taken, and then be discovered to be untrue with the result that the liar suffers some sort of punishment, as in the case of prosecutions for fraud. This would require a somewhat more complicated set of probabilities than the ones I have specified.
7. The possibility of miscarriages of justice should be taken into account if this line of reasoning is further extended. Since I am working on a book on the general subject of the law, I will put the matter aside in this essay.
8. Gordon Tullock, *The Organization of Inquiry* (Durham, N.C.: Duke University Press, 1967).

Notes to Chapter X

1. The size of the districts used by countries following European types of PR varies a good deal. In the case of Israel, it is the entire country. Most of the others use varying sizes running from five to thirty. One of the minor advantages of the system is that it makes redistricting, to take care of shifts of population, relatively easy. A seat can be added or subtracted if changing boundaries is inconvenient.
2. Charging a fee for being hooked up to the electronic voting network and for the broadcast of the proceedings of the assembly might further reduce the number of people who chose to vote on measures in spite of very small popular support.
3. James M. Burns, *The Deadlock of Democracy* (Englewood Cliffs, N.J.: Prentice-Hall, Inc., 1963).
4. As an exception to this, some prominent political scientists who were basically partisans of the Democratic party suggested that Truman appoint a Republican Secretary of State and resign after the 1946 election.
5. This is not always so. In William Riker's *The Theory of Political Coalitions,* it is demonstrated that the power of a clique engaged in negotiation to form a majority coalition is not a smooth function of its size. In general, however, the larger the party, the more powerful.
6. See James Buchanan and Gordon Tullock, *The Calculus of Consent*

(Ann Arbor: The University of Michigan Press, 1962), and Gordon Tullock, *Entrepreneurial Politics* (Thomas Jefferson Center, University of Virginia, 1962).

7. *American Political Science Review* (June 1961), p. 368.
8. Teachers normally have a roughly similar problem in dealing with questions from the class. As a general rule, the number of questions asked by a class of thirty will not be markedly different from the number asked by a class of ten. Further, it may well be the case that in each case almost all of the questions come from two or three of the students.
9. Except for the spectators. Teachers of political science who bring their students to Washington to see how the nation is governed normally get quite disturbed by the sight.
10. It may be thought unwise to give strong points of view so much time. If so, the computer could be instructed to make the appropriate adjustments in the points allotted to to the members of different points of view.
11. It would be necessary to introduce some arbitrary rule to get the system started and to break ties. It might be wise to permit members to get priority so that they could speak out of turn if they agreed to "pay" twice the normal rate in points.

Index

Selected Ann Arbor Paperbacks
Works of enduring merit

AA 4 **THE SOUTHERN FRONTIER, 1670-1732** Verner W. Crane
AA 9 **STONEWALL JACKSON** Allen Tate
AA 13 **THOMAS JEFFERSON: The Apostle of Americanism** Gilbert Chinard
AA 18 **THOMAS MORE** R. W. Chambers
AA 21 **THE PURITAN MIND** Herbert W. Schneider
AA 28 **ANTISLAVERY ORIGINS OF THE CIVIL WAR IN THE UNITED STATES** Dwight Lowell Dumond
AA 31 **POPULATION: THE FIRST ESSAY** Thomas R. Malthus
AA 34 **THE ORIGIN OF RUSSIAN COMMUNISM** Nicolas Berdyaev
AA 35 **THE LIFE OF CHARLEMAGNE** Einhard
AA 49 **THE GATEWAY TO THE MIDDLE AGES: ITALY** Eleanor Shipley Duckett
AA 50 **THE GATEWAY TO THE MIDDLE AGES: FRANCE AND BRITAIN** Eleanor Shipley Duckett
AA 51 **THE GATEWAY TO THE MIDDLE AGES: MONASTICISM** Eleanor Shipley Duckett
AA 53 **VOICES OF THE INDUSTRIAL REVOLUTION** John Bowditch and Clement Ramsland, ed.
AA 54 **HOBBES** Sir Leslie Stephen
AA 55 **THE RUSSIAN REVOLUTION** Nicolas Berdyaev
AA 56 **TERRORISM AND COMMUNISM** Leon Trotsky
AA 57 **THE RUSSIAN REVOLUTION and LENINISM OR MARXISM?** Rosa Luxemburg
AA 59 **THE FATE OF MAN IN THE MODERN WORLD** Nicolas Berdyaev
AA 61 **THE REFORMATION OF THE 16TH CENTURY** Rev. Charles Beard
AA 62 **A HISTORY OF BUSINESS: From Babylon to the Monopolists Vol. I** Miriam Beard
AA 65 **A PREFACE TO POLITICS** Walter Lippmann
AA 66 **FROM HEGEL TO MARX: Studies in the Intellectual Development of Karl Marx** Sidney Hook
AA 67 **WORLD COMMUNISM: A History of the Communist International** F. Borkenau
AA 69 **THE MYTH OF THE RULING CLASS: Gaetano Mosca and the Elite** James H. Meisel
AA 72 **THE MERCHANT CLASS OF MEDIEVAL LONDON** Sylvia L. Thrupp
AA 74 **CAPITALISM IN AMSTERDAM IN THE 17TH CENTURY** Violet Barbour
AA 76 **A HISTORY OF BUSINESS: From the Monopolists to the Organization Man Vol. II** M. Beard
AA 77 **THE SPANISH COCKPIT** Franz Borkenau
AA 78 **THE HERO IN AMERICA** Dixon Wecter
AA 79 **THUCYDIDES** John H. Finley, Jr.
AA 80 **SECRET HISTORY** Procopius
AA 86 **LAISSEZ FAIRE AND THE GENERAL-WELFARE STATE** Sidney Fine
AA 88 **ROMAN POLITICAL IDEAS AND PRACTICE** F. E. Adcock
AA 94 **POETRY AND POLITICS UNDER THE STUARTS** C. V. Wedgwood
AA 95 **ANABASIS: The March Up Country** Xenophon Translated by W. H. D. Rouse
AA 100 **THE CALCULUS OF CONSENT** James M. Buchanan and Gordon Tullock
AA 103 **IMPERIALISM** J. A. Hobson
AA 104 **REFLECTIONS OF A RUSSIAN STATESMAN** Konstantin P. Pobedonostsev
AA 110 **BAROQUE TIMES IN OLD MEXICO** Irving A. Leonard
AA 111 **THE AGE OF ATTILA** C. D. Gordon
AA 114 **IMPERIAL GERMANY AND THE INDUSTRIAL REVOLUTION** Thorstein Veblen
AA 115 **CIVIL LIBERTIES AND THE CONSTITUTION** Paul G. Kauper
AA 118 **NEGRO THOUGHT IN AMERICA** August Meier
AA 119 **THE POLITICAL IDEAS OF THE ENGLISH ROMANTICISTS** Crane Brinton
AA 120 **WILLIAM PENN** Catherine Owens Peare
AA 122 **JOAN OF ARC** Jules Michelet Translated by Albert Guérard
AA 124 **SEEDTIME OF REFORM** Clarke A. Chambers
AA 126 **LECTURES ON THE PRINCIPLES OF POLITICAL OBLIGATION** T. H. Green
AA 133 **IMPRESSIONS OF LENIN** Angelica Balabanoff
AA 137 **POLITICAL HERETICS** Max Nomad
AA 139 **REBEL VOICES: An I.W.W. Anthology** Joyce L. Kornbluh, ed.
AA 160 **PATTERNS OF SOVIET THOUGHT** Richard T. De George
AA 172 **DEATH AND LIFE IN THE TENTH CENTURY** Eleanor Shipley Duckett
AA 173 **GALILEO, SCIENCE AND THE CHURCH** Jerome J. Langford
AA 179 **SEARCH FOR A PLACE** M. R. Delany and Robert Campbell
AA 181 **THE DEVELOPMENT OF PHYSICAL THEORY IN THE MIDDLE AGES** James A. Weisheipl
AA 186 **CAROLINGIAN CHRONICLES** Translated by Bernhard Walter Scholz with Barbara Rogers
AA 187 **TOWARD A MATHEMATICS OF POLITICS** Gordon Tullock

For a complete list of Ann Arbor Paperback titles write:
THE UNIVERSITY OF MICHIGAN PRESS ANN ARBOR